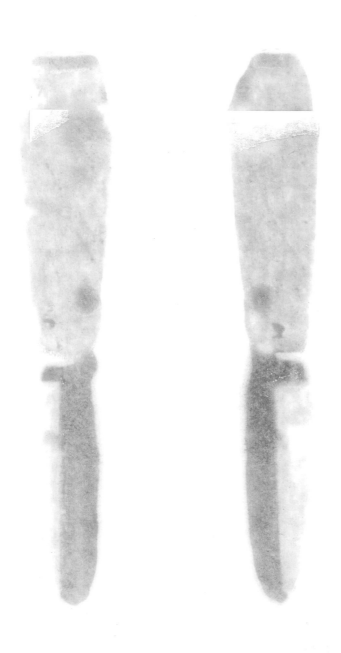

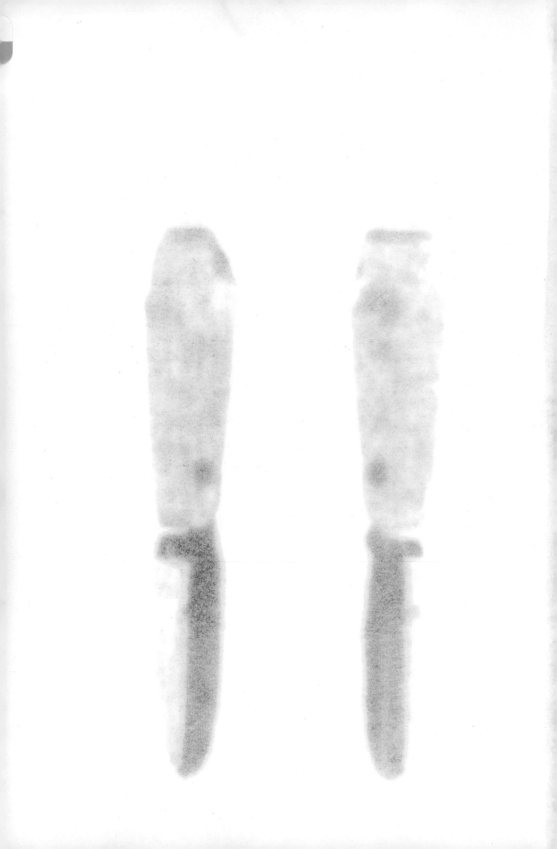

PCB POISONING AND POLLUTION

PCB POISONING
AND
POLLUTION

Edited by

Kentaro HIGUCHI

Fukuoka University Hospital, Japan

1976

 KODANSHA LTD.
Tokyo

 ACADEMIC PRESS
New York   San Francisco   London

A Subsidiary of Harcourt Brace Jovanovich, Publishers

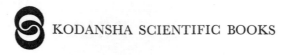

KODANSHA SCIENTIFIC BOOKS

Copyright © 1976 by Kodansha Ltd.

I.S.B.N. 0-12-347850-2
Library of Congress Catalog Card Number 76-24154

Co-published by
KODANSHA LTD.
12-21, Otowa 2-chome, Bunkyo-ku, Tokyo 112, Japan
and
ACADEMIC PRESS, INC.
111 Fifth Avenue, New York, N.Y. 10003
ACADEMIC PRESS, (LONDON) LTD.
24/28 Oval Road, London NW 1

PRINTED IN JAPAN

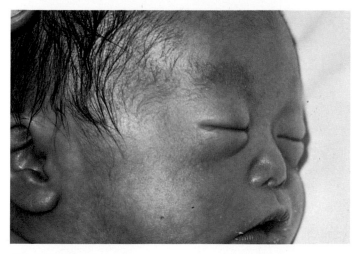

Fig. 7.15 Swelling of the eyelids and brownish skin pigmentation of a new-born baby.

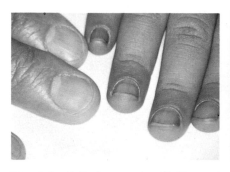

Fig. 7.9 Nail pigmentation (S. K., aged 5, female).

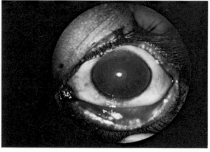

Fig. 7.10 Pigmentation of the palpebral conjunctiva and limbal conjunctiva.

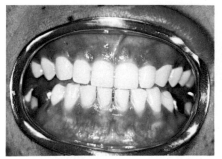

Fig. 7.11 Gingival pigmentation (N. K., aged 36, female).

Fig. 7.12 Pigmentation of the lips.

Contributors

Kentaro HIGUCHI, Fukuoka University Hospital, Nishi-ku, Fukuoka 814, Japan

Chisato HIRAYAMA, Third Department of Internal Medicine, Faculty of Medicine, Kyushu University, Fukuoka 812, Japan.

Masahiro KIKUCHI, Department of Pathology, Faculty of Medicine, Fukuoka University, Fukuoka 814, Japan

Hiromu KODA, Department of Dermatology, Faculty of Medicine, Kyushu University, Fukuoka 812, Japan

Masanori KURATSUNE, Department of Public Heath, Faculty of Medicine, Kyushu University, Fukuoka 812, Japan

Yoshito MASUDA, Daiichi College of Pharmaceutical Sciences, Fukuoka 815, Japan

Hiroya TANABE, Food Hygiene Laboratory, Sagami Women's University, Sagamihara-shi 228, Japan

Roy TATSUKAWA, College of Agriculture, Ehime University, Matsuyama-shi 790, Japan

Kiichi UEDA, Laboratory of Hygiene and Oral Health, Showa University, Tokyo 142, Japan

Harukuni URABE, Department of Dermatology, Faculty of Medicine, Kyushu University, Fukuoka 812, Japan

Shin'ichi YOSHIHARA, Faculty of Pharmaceutical Sciences, Kyushu University, Fukuoka 812, Japan

Hidetoshi YOSHIMURA, Faculty of Pharmaceutical Sciences, Kyushu University, Fukuoka 812, Japan

Preface

Recently, the various health hazards and dangers of indiscriminate discharge of industrial wastes into the environment have been emphasized. Also, strong warnings have been issued, both officially and unofficially, against polluting the environment with toxic or otherwise harmful industrial products. However, it is distressing that, so often, severe cases of poisoning are required before the dangers are fully recognized or effective countermeasures can be implemented or enforced.

The present book covers the broad field of PCB poisoning and pollution in Japan. It emphasized the case of direct human poisoning by oral intake of PCB-contaminated rice-oil (Yusho), and also includes sections on the experimental and toxicological aspects of PCB poisoning, and on the general pollution of the natural environment by PCB's.

The medical description of Yusho covers epidemiologic, pathologic, clinical and therapeutic studies, and there are detailed descriptions of the dermal symptoms (which resemble chloracne) and other major clinical manifestations of the disease, in both adults and children. The sections on toxicology cover various studies, mostly animal experiments, on the physiological and other effects of PCB's, and on the absorption, distribution and metabolism of PCB's in the animal body. The sections devoted to environmental pollution in Japan describe both the microanalytical methodology for PCB's, as adopted in Japan, and also the extent of pollution of the atmosphere, freshwater and marine environments, soil, agricultural products, fish, birds, and man.

As editor for this project, I wish to express my appreciation to all authors for their contributions, and would also like to take this opportunity to thank the staff of Kodansha for their editorial and linguistic assistance in the preparation of the English manuscripts comprising this book.

July, 1976

K. Higuchi

Contents

PCB POISONING

1

Outline

Kentaro HIGUCHI

On June 7, 1968, a 3-yr-old girl came to the Out-patient Clinic, Dept. of Dermatology, Kyushu University, suffering mainly from an acneform skin eruption. Her parents and elder sister also developed similar symptoms and were given medical attention in early August. The number of such patients increased steadily and the disease was soon found to be widely distributed in northern Kyushu, especially Fukuoka Pref. Clinical dermal findings resembled those of chloracne, and so it was suspected that the disease might be due to poisoning by organic chlorine or some agricultural chemical. However, it was necessary first to establish the basic symptomatology and to investigate the problems related to familial outbreaks and differences in the patients' life environment.

The first detailed report elucidated the basic symptomatology of the disease, utilizing data accumulated from clinical and laboratory examinations of 138 patients visiting Kyushu University Hospital as out-patients between June 1968 and January 1969. The essential findings at that time may be summarized as follows:

NO. OF PATIENTS: 138 (male 72, female 66).
AGE: under 10 (34 cases), 10–19 (20 cases), 20–29 (27 cases), 30–39 (29 cases), 40 and over (less common by 10-yr groups; total, 28 cases).

3

SEX: under-10 group (boys 22, girls 12); adults (no significant difference by sex).

TIME OF OUTBREAK: March 1968 (8 cases), April (13 cases), May (20 cases), June (32 cases), July (18 cases), August (28 cases), September (10 cases), October (2 cases), November (no cases), December (5 cases), January 1969 (2 cases). Most of the 28 cases in August were children under 10. The data thus suggested that the major outbreak was between May and August, centering on June.

SYMPTOMS OF THE INCIPIENT STAGE: increased discharge from the eye, swelling of the eyelids, and weakness of eyesight (52 cases); acneform eruptions and enlargement of the follicular openings (45 cases); pigmentation of the nails, gingivae and lips (13 cases); swelling of the limbs (9 cases); nausea and vomiting (4 cases). Ocular symptoms generally appeared first, followed 2–3 months later by dermal symptoms; general effects such as fatigue, nausea and vomiting were also early.

CLINICAL SIGNS: many patients complained of swelling of the upper eyelids, anorexia, languishment, a feeling of numbness, edematous swelling of the limbs, nausea, vomiting, lumbago, joint pains, heel pains, pigmentation of the lips, abnormal menstruation, and impotence. Various symptoms of chloracne were thus present, associated with follicular keratosis.
(1)   *Ophthalmic findings*: unevenness of line in the eyelids, swelling of the meibomian glands, pigmentation of the corneal ring and conjunctiva, hypersecretion of cheese-like material from the eye.
(2)   *Dermatological findings*: acneform eruptions, marked enlargement of the follicular openings, pigmentation and flattening of the nails, pigmentation of the skin, lips, gingivae and mucous membrane of the oral cavity, hyperkeratotic plaque formation in the soles and palms, hyperidrosis, hypertrichosis, swelling of Montgomery's glands, cyst formation in genital sebaceous glands, and (especially in children) dry skin and miliaria at joint sites.
   The dermal changes appeared to be based on hyperkeratinization associated with abnormal lipid metabolism. Individual lesions of the acneform eruptions ranged from pinhead- to pea-size. They constituted pale straw-colored cysts, sometimes in the form of large eruptions filled with keratinous but little sebaceous material. The hair follicles exhibited enlargement of the follicular opening and had a filling of keratinous material. There was considerable secondary infection by *Staphylococcus epidermidis*, yielding large abscesses and lesions such as infectious atheromas. Some cases of hemispheric tumors in the joints were also observed.

GRADING ACCORDING TO SEVERITY OF DERMAL SYMPTOMS: GRADE I, cheese-like discharge from swollen meibomian glands, and nail pigmentation;

GRADE II, as I, but with comedo formation; GRADE III, as II, but with clear acneform eruptions, and cysts in the genitocrural region; GRADE IV, as III, but with enlargement of the follicular openings over the entire body surface and more general occurrence of acneform eruptions. Light grades were predominant in children, and high grades predominant between adolescence and 40.

LABORATORY FINDINGS: Laboratory tests showed increased serum alkaline phosphatase, and rises in SGOT and SGPT values in severe cases. Elevated serum triglyceride levels were observed in general, and serum cholesterol was low in serious cases. Serum protein analysis indicated rises in albumin and α_2-globulin, and a fall in γ-globulin. Such findings appeared to be characteristic of the disease, and contrasted with the normal serum alkaline phosphatase and SGOT and SGPT values, increased serum cholesterol and total lipids, and normal analytical percentages of serum proteins of occupational acne.

Based on published reports, the first case of occupational chloracne is that described by Herxheimer from a German factory in 1899.[1] Similar symptoms due to chlornaphthalene ("Perchlornaphthalin", a kind of wax formerly used as an insulator of electric cells) were also encountered during World War II, giving rise to the name "Pernakrankheit". Reports of such occupational chloracne subsequently followed from various chemical factories, etc.

In our case, however, the observed chloracne-like disease apparently had no relation to any industrial occupation, since it occurred familially in the general home. To determine its cause, a Study Group was organized in October 1968, comprising clinical survey and chemical analysis members from the Faculties of Medicine, Pharmaceutical Sciences and Technology, Kyushu University. As a result of much laborious research, it became apparent that the causative agent was probably mixed with rice-bran oil, a common element in the diet of affected families. This hypothesis was soon confirmed by the finding that affected family units were limited to those using a very particular brand of rice oil. Analytical investigations then showed specific lots of this oil to be contaminated with Kanechlor 400 (PCB's), the main component of which was tetrachlorobiphenyl. The disease was thus named "Yusho" (lit. oil disease).

The only previous report of chloracne induced by oral intake of a toxic substance was that of Herzberg,[2] who described a case of fried-potato poisoning due to chlorparaffin intake. Yusho was thus apparently the first instance of any large-scale outbreak of this nature.

Suspect oil samples were collected, registered and kept under the supervision of the Dept. of Legal Medicine, to serve for subsequent chemical analysis and animal experiments. Diagnostic criteria and a treatment guide for Yusho were published and distributed, and medical examinations and field surveys were performed by group members in close cooperation with the public health centers.

The findings obtained by late February, 1969, may be summarized as follows:

1)   Occurrence of the disease was limited to family units which used Kanemi-brand rice oil produced in early February 1968.

2)   Kanechlors (chlorinated diphenyls or chlorobiphenyls) which were used in the manufacturing process, were present in the Kanemi rice oil used by the patients, and *ca.* 2000 ppm Kanechlor 400 was detected in oil taken by the patients.   The total amount of oil ingested by one patient's family (av. 4 people) was about 1.8 l/month.   The peroxide value of the oil (as determined by the Jod-tit method) was 27.2 me$_q$/kg or more, in contrast to 3.1–5.9 me$_q$/kg for general foodstuffs.   This oil contained about 37% of C18–2 (linoleic acid).

3)   Chemical analysis of samples from Yusho patients showed Kanechlor levels of 13.1–75.5 ppm in their sebum and subcutaneous fat.

4)   It was satisfactorily concluded that the disease was the result of chlorobiphenyl poisoning, and that this was induced by the ingestion of rice oil contaminated with Kanechlor 400, a PCB mixture used for minimizing rice-oil odor during the manufacturing process.

It should be mentioned also that, just before the Yusho outbreak (i.e. in February–March, 1968), an instance of accidental mass death of chickens had occurred in Japan.   Its cause was found to be a dark oil used in the chicken feed, which analysis also confirmed to contain Kanechlor 400.   This case thus resembles that occurring in the eastern U.S.A. in 1957, when chicken edemas were found.[3]

In subsequent work on Yusho, the Study Group obtained successful results in the following fields:

1)   Important clinical observations were made on otorhinolaryngological, respiratory, oral, ocular, gynecological and neurological effects, in addition to the dermatological work.

2)   Clinico-chemical investigations were continued and expanded in the hematological and serological fields, particularly in relation to hypertriglyceridemia of the blood.

3)   Detailed histological changes in the skin and liver were identified based on samples from patients, stillborns and adult autopsy cases.

4)   Experimental work on the metabolic effects of orally administered Kanechlor 400 in rats and hairless mice was carried out.   This included studies on the tissue distribution of PCB's, the content of different Kanechlor 400 components in the blood, sputum, etc., and the urinary and fecal excretion of PCB's in mice.

5)   Surveys of pregnant women who had consumed contaminated rice oil, their babies, and on deviations from normal sexual function in female Yusho patients were undertaken.

6)   Detailed epidemiologic studies were made for a total of over 1000 cases, including about 20 mortality cases.

7)   Therapeutic trials were undertaken with various medicines, and by balneotherapy and fasting therapy, although the degree of success was at best limited.

In addition, it should be added that other instances of PCB pollution subsequently occurred, including that by noncarbon copy paper, etc.

The remaining chapters of this section consider the detailed aspects of Yusho and related problems, and in the case of Yusho itself, general reviews of the work of the Study Group have previously appeared in *Fukuoka Acta Medica* as Report No. 1 (1969),[4] No. 2 (1971)[5] and No. 3 (1972).[6]

REFERENCES

1.   K. Herxheimer, *Münch. Med. Wschr.*, **46**, 178 (1899).
2.   H. J. Herzberg, *Derm. Wschr.*, **7**, 425 (1947).
3.   D. F. Flick, R. G. O'Dell and V. A. Childs, *Poultry Sci.*, **44**, 1460 (1965).
4.   Reports on the Study of "Yusho" (Chlorobiphenyl Poisoning), No. 1, *Fukuoka Acta Med.*, **60**, 403 (1969).
5.   *Ibid.*, No. 2, *Fukuoka Acta Med.*, **62**, 1 (1971).
6.   *Ibid.*, No. 3, *Fukuoka Acta Med.*, **63**, 347 (1972).

2

Epidemiologic Studies on Yusho

Masanori KURATSUNE

2.1 INTRODUCTION

As described in Chapter 1, an epidemic of a peculiar disease similar to chloracne was first encountered in 1968 in Fukuoka Pref., Japan. The epidemic was later found to have spread not only across Fukuoka Pref. but also to about 20 other prefectures in the western part of the country (Fig. 2.1). It produced a total of 1200 patients, according to the Government tabulation of September, 1973.[1]

As members of the Study Group on the epidemic, the author and his associates conducted extensive epidemiologic investigations following the basic methodology of epidemiology.[2] Fortunately, the cause of the disease was soon identified by the Study Group to be the ingestion of rice oil contaminated with PCB's, and the disease was named "Yusho" (lit. oil disease). Although our direct observations and experience have been confined to patients occurring in Fukuoka Pref., it is considered that they are also applicable in general to the Yusho patients seen in other prefectures.

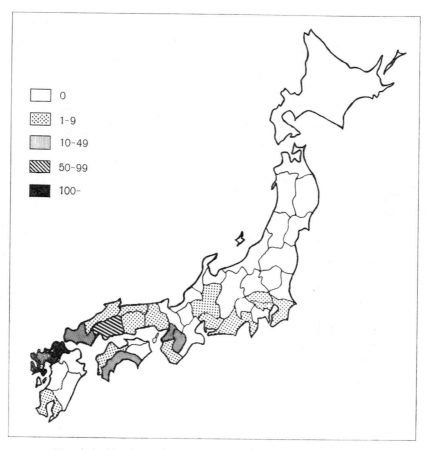

Fig. 2.1   Numbers of Yusho patients by prefecture.

2.2   Clinical Symptoms

The most common symptoms experienced by 136 early Yusho patients at the incipient stage[3] were increased discharge from the eyes and swelling of the upper eyelids (38.2%), followed by acneform eruptions and follicular accentuation (33.1%), and skin pigmentation (9.6%). The subjective symptoms of Yusho, as indicated by 189 patients diagnosed up to October 31, 1968, are listed in Table 2.1.[2]

TABLE 2.1

Percentage Distribution of Yusho Symptoms as Reported by
189 Patients Examined before October 31, 1968

Symptoms	Males (N = 89)	Females (N = 100)
Dark-brown nail pigmentation	83.1	75.0
Distinctive hair follicles	64.0	56.0
Increased sweating at the palms	50.6	55.0
Acneform skin eruptions	87.6	82.0
Red plaques on the limbs	20.2	16.0
Itching	42.7	52.0
Pigmentation of the skin	75.3	72.0
Swelling of the limbs	20.2	41.0
Stiffened soles of the feet and palms of the hands	24.7	29.0
Pigmented mucous membrane	56.2	47.0
Increased eye discharge	88.8	83.0
Hyperemia of the conjunctiva	70.8	71.0
Transient visual disturbance	56.2	55.0
Jaundice	11.2	11.0
Swelling of the upper eyelids	71.9	74.0
Feeling of weakness	58.4	52.0
Numbness of the limbs	32.6	39.0
Fever	16.9	19.0
Hearing Difficulties	18.0	19.0
Spasms of the limbs	7.9	8.0
Headaches	30.3	39.0
Vomiting	23.6	28.0
Diarrhea	19.1	17.0

2.3   DESCRIPTIVE EPIDEMIOLOGIC STUDIES

Initially, 325 patients encountered in Fukuoka Pref. from October, 1968 to January, 1969 were analyzed in order to determine the distributional characteristics of the disease.   One of the most important features immediately noted was a distinct familial aggregation.   The 325 patients belonged to only 112 families.   Also, as shown in Fig. 2.2, about 98% of the patients were affected between February and October, 1968, while 4 patients (1.2%) stated that they had become ill as early as December, 1967.   About 55% of the patients were concentrated in the 3 months from June to August and no significant differences were found between the sexes in the monthly distributions.

In order to examine the geographical distribution of patients, crude incidence rates of the disease were calculated for 3 large cities and for the jurisdictional areas of the 22 local health departments in Fukuoka Pref.   The observed rates varied considerably from 0.0 to 58.9 per 100,000.   Such examinations of the geographical distribution failed to reveal any

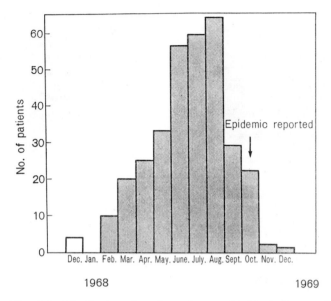

Fig. 2.2 Distribution of patients by month of appearance of Yusho symptoms.

common socioeconomic or environmental factor which might be associated with the disease.

The 325 patients comprised 158 males and 167 females, indicating that both sexes were affected equally. Age- and sex-specific incidence rates again showed no significant difference by sex, and also revealed that all age groups were affected, although a lower risk was apparent for both males and females over 60 yr old (Table 2.2).

TABLE 2.2
Age-specific Incidence Rates for Yusho, by Sex of Patients

Age group (yr)	Incidence rates per 100,000	
	Males	Females
0— 9	11.4	8.6
10—19	8.9	6.6
20—29	9.1	10.3
30—39	9.6	11.9
40—49	5.4	9.3
50—59	5.4	7.2
60—69	3.5	2.4
70 and over	1.7	0.0
Total	8.3	8.1

2.4   ANALYTICAL EPIDEMIOLOGIC STUDIES

When our study started, a commercial brand of rice oil produced by Kanemi Sōko Co. (abbreviated here as "K rice oil") was suspected as a possible cause of the disease since most of the Yusho patients appeared to have used it.   In order to examine this suspicion, thorough investigations were undertaken with regard to the lot numbers appearing on the left-over containers of the oil used by patients, the shipping records of K company, and the purchase and sales records of wholesale oil dealers and retail stores.   It was soon discovered that all the patients had used K rice oil at some time, either in canned form (16.5 kg) or bottled form (1.65 kg). Furthermore, an astonishing fact became evident: regardless of where they lived, 166 of 170 patients who had used only canned K rice oil had used a very specific oil, viz. that produced or shipped by the company on February 5 and 6, 1968 (Table 2.3).   For those using bottled rice oil, the

TABLE 2.3
Details of Rice Oil Used by Yusho Patients

No. of Patients (%)	Specifications of oil used	
	Type	Date of production or shipment
166 (51.1)	Canned	February 5 and 6, 1968
4 (1.2)	Canned	Unknown
143 (44.0)	Bottled	Unknown, but use of oil shipped between February 5 and 15, 1968, is very probable.
12 (3.7)	Bottled	Unknown
325 (100.0)		

dates of production or shipment could not be confirmed, since they had not retained the old bottles at home.   However, an examination of all available shipping, sales, and purchase records of K company, and of wholesale dealers and retail stores, clearly suggested a possibility that at least 143 of 155 patients who had used bottled K rice oil only had used a very specific oil, viz. that produced or shipped between February 5 and 15, 1968.   This was based on such oil having been shipped to and having reached the retail stores from which the patients had purchased their oil. Thus, almost all the patients had undergone the very peculiar common experience of using, or possibly using, rice oil produced or shipped by a single company over a specific period of time (Table 2.3).   Moreover, an attack rate of as high as 63.9% was apparent among those who had consumed the specific oil in canned form.

In an additional survey, we examined whether or not those who had regularly used K rice oil, but had not used the specific K rice oil produced or shipped in the period in question, were free of the disease.   A group of 113 persons from 29 households living in an apartment block were found to have purchased canned K rice oil as a unit from one dealer, dividing it among themselves fairly regularly from December, 1967 to September, 1968 (except for the period January to April, 1968, when no bulk purchases were made).   Their disease experience in 1968 was carefully examined through an investigation of their medical records at hospitals and doctors' offices, and no case of Yusho was found among them.

Although these results all strongly suggested that Yusho was indeed caused by intake of the very specific K rice oil, there still remained some possibility that certain other factors or agents might be involved as a primary or secondary cause of the epidemic.   Therefore, two case-control studies were undertaken.   In one, 121 patients and their randomly selected 121 healthy controls (53 males and 68 females, matched individually by age, sex and place of residence to the patients) were asked 60 questions concerning their occupation, medical background, general health status, habits, customs, diet, pets, and other characteristics of their daily lives. As shown in Table 2.4, only one of the 60 personal factors examined, namely the habit of "eating fried foods or tempura almost every day",

TABLE 2.4

Results of a Case-Control Study on Daily Life

Items investigated	Cases (%)	Controls (%)
Allergic to fish	5.0	7.5
Allergic to aspirin	0.0	4.2
Allergic to other drugs	7.5	6.6
Having bath facilities at home	84.7	85.5
Taking a bath every day	73.0	70.6
Having pets at Home	18.3†	36.5†
Living in a house smaller than 66 m² floor space	66.9	66.1
Handling agricultural chemicals	2.5	6.6
Taking cod liver oil	10.8	8.3
Taking vitamin pills	23.2	18.3
Taking other restorative drugs	9.1	7.5
Water supply available at home	81.3	74.7
Dining out occasionally	28.1	30.6
Dining on the same meals as family	88.8	89.6
Eating green vegetables daily	63.1	58.9
Drinking milk nearly every day	49.0	39.0
Taking butter nearly every day	22.4	24.9
Eating eggs nearly every day	64.7	59.8
Eating fried foods or tempura nearly every day	22.4†	11.6†
Eating foods prepared with oil nearly every day	45.7	35.7
Eating fish nearly every day	21.6	29.1
Taking mayonnaise nearly every day	10.8	10.8
Eating instant "ramen" or chinese noodles nearly every day	10.8	10.0

† $p < 0.05$.

TABLE 2.5
Results of a Case-Control Study on Oils Used

Fat or oil	Cases[1]		Controls[2]	
	No. of households	(%)	No. of households	(%)
Butter	35	50.7	105	50.7
Margarine	44	63.8	127	61.4
Sesame oil	21	30.5	85	41.1
Rapeseed oil	10	14.5	77	37.2
Rice bran oil	66	95.7	64	30.9
Lard	12	17.4	38	18.4
Other oils	13	18.8	117	56.5

[1] Cases: 69 patient households
[2] Controls: 207 non-patient households

was significantly more common among the patients than among the controls.

In the second case-control study, 69 of the households with Yusho patients were matched by place of residence with 207 control households without such patients. A distinct difference was noted between the two groups only as regards their regular use of rice oil (Table 2.5). Thus, the case-control studies clearly indicated that none of the factors tested, apart from the use of rice oil, could satisfactorily account for the disease.

2.5   DOSE-RESPONSE RELATIONSHIP

In order to prove this apparent causal relationship, a definite dose-response relationship was needed. A rough estimate of the quantity of the specific K rice oil consumed by each patient and his family members was made, disregarding age, sex, amount of food intake, and possible losses of oil during and after cooking.[4] As shown in Table 2.6, 80 of 146 users of the specific K rice oil were believed to have consumed, individually, less than 720 ml. For these 80 light users, the attack rate of Yusho was 88%, whereas for those estimated to have used more than 720 ml, it was 100%. Furthermore, possible relationships between amount of oil consumed and clinical severity of the patients were examined. It was found that while the clinical severity of the disease did not differ signifi-

TABLE 2.6
Relation between Amount of K Rice Oil Used by Patients
and Their Clinical Severity

Amount of oil used (ml)	Non-affected		Light cases		Severe cases		Total	
	No.	(%)	No.	(%)	No.	(%)	No.	(%)
<720	10	(12.0)	39	(49.0)	31	(39.0)	80	(100.0)
720—1439	0	(0.0)	14	(31.0)	31	(69.0)	45	(100.0)
>1440	0	(0.0)	3	(14.0)	18	(86.0)	21	(100.0)

TABLE 2.7
Clinical Severity by Age and Amount of Oil Used

Amount of oil used (m*l*)	Age (yr)	No. of patients by clinical grade (males and females)							
		Non-affected		Light†		Severe†		Total	
<720	0—12	3	(16.7)	13	(72.2)	2	(11.1)	18	(100.0)
	13—29	2	(8.3)	7	(29.2)	15	(62.5)	24	(100.0)
	30	5	(13.1)	19	(50.0)	14	(36.9)	38	(100.0)
	Total	10	(12.5)	39	(48.8)	31	(38.7)	80	(100.0)
720—1439	0—12	0	(0)	5	(50.0)	5	(50.0)	10	(100.0)
	13—29	0	(0)	1	(6.7)	14	(93.3)	15	(100.0)
	30—	0	(0)	8	(40.0)	12	(60.0)	20	(100.0)
	Total	0	(0)	14	(31.2)	31	(68.8)	45	(100.0)

† "Light" corresponds to GRADES I and II, and "severe" to GRADES III and IV of Goto *et al.*, (see also chapter 7).

cantly between the sexes, it varied considerably according to age (Table 2.7). The proportion of severe cases among those aged 13 to 29 was significantly higher than that of other age-groups. Each of the three groups of users with different levels of oil intake was therefore standardized for age, using the age composition of the complete group of 146 users as standard. The resulting figures, however, were largely unchanged from those of Table 2.6, demonstrating that the proportion of severe cases of Yusho increased significantly with the amount of oil consumed. Thus, a clear dose-response relationship was demonstrated, although the dose estimates were of course approximate.

2.6 TOXIC AGENT

All these epidemiologic findings thus led us to conclude that the K rice oil of specific production or shipment had caused Yusho. However, this conclusion needed to be verified by elucidating the reason why the oil was toxic. The chemical study subgroup of the Sutdy Group on Yusho found that the canned K rice oil produced or shipped on February 5, 1968, and used by some of the patients, contained about 2000–3000 ppm of Kanechlor 400 (a brand of polychlorinated biphenyls of chlorine content 48% produced by Kanegafuchi Chemical Industry Co.). It was also demonstrated that the oil was not contaminated with other toxic agents such as Cu, Ni, Zn, Co, As, Hg, or pentachlorophenol. Furthermore, most of the components of Kanechlor 400, particularly those corresponding to peaks of higher retention time in gas-chromatograms, were found in the sebum, subcutaneous fat, mesenterium, mesenteriolum, extraperitoneal adipose tissue, appendix vermiformis, heart, sternal marrow, small intestine, trachea and other organs of patients, and also in fetal tissues.[5–7]

In order to test whether it was only the K rice oil produced or shipped during the specific period in question that was contaminated with Kane-chlor, 109 samples of bottled rice oil shipped between October, 1967, and October, 1968, were analyzed.   Gas-chromatographic analysis showed that, for the bottled oil, contamination was limited to material produced or shipped between February 7 and 19, 1968, although traces of contamination were also noted in a few samples produced in mid-March, 1968.[5] No analysis could be made for bottled K rice oil produced or shipped on February 5 and 6, 1968, due to a lack of samples.   Similarly, 479 random samples of bottled K rice oil were analyzed for their chlorine content by the X-ray fluorescence method using a count meter.   Again, only samples from the middle of February were found to contain significant amounts of chlorine (max. 462 ppm, in the February 7 sample, followed by a quick decrease).   None of the oil shipped in other months was contaminated with more than a trace of chlorine.   Thus, the results of the chemical studies were completely consistent with those obtained by the epidemiologic approach.

Finally, in this connection, it should be noted that Kanechlor had been used by K company in its equipment for heating the processed oil at reduced pressure in order to remove odorous matter from the oil (Fig.

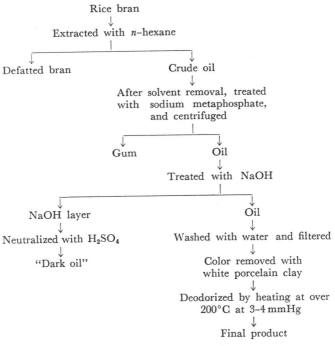

Fig. 2.3   Flow sheet for rice oil production.

2.3).   It is therefore considered that Kanechlor must have leaked from the heating pipe and contaminated the oil.   This conclusion is supported by the discovery of small holes in an old section of the pipe.

Recently, we analyzed various types of Kanechlor and 3 samples of the toxic rice-oil used by Yusho patients for polychlorinated dibenzofurans (PCDF) and polychlorinated dibenzo-p-dioxins (PCDD), using a gas chromatograph and a mass spectrometer.   Kanechlor-300, -400, -500, and -600 were found to contain 1, 18, 4, and 5 ppm of PCDF, as calculated from the gas-chromatographic peak heights, or 1.5, 17, 2.5, and 3 ppm as determined by the perchlorination method respectively.   On the other hand, the samples of toxic rice-oil contained 800–1000 ppm of PCB's and 5 ppm of PCDF, consisting mainly of tetra- and pentachlorodibenzofurans. Thus, the relative concentration of PCDF to the PCB in these oils was for unknown reasons much higher than expected from the known concentration of PCDF in Kanechlor-400.   The possibility that PCDF played a role in Yusho should therefore not be dismissed.   PCDD was not shown to be present in either the oils or any of the other materials tested.

2.7   Amount of Kanechlor 400 Ingested by Patients

Among the patients, 146 who were known to have used the contaminated canned K rice oil of February 5 and 6 were selected for estimation of the total amount of Kanechlor 400 ingested.   As mentioned above, it was possible to calculate only the approximate amount of the specific oil consumed by each of the patients, giving an average of about 800 ml. Since the concentration of Kanechlor 400 in the oil was 2000–3000 ppm, the average amount of Kanechlor 400 ingested by the patients could be estimated roughly at 2 g.[4]   Similarly, the minimum amount of Kanechlor 400 consumed by the patients was estimated at about 0.5 g.   However, caution should be exercised in interpreting these figures, on at least three counts.   First, the patients did not actually ingest Kanechlor 400 as such.   What they consumed was rice oil contaminated with "used" Kanechlor 400.   Certain of the components of Kanechlor 400 were undoubtedly missing from the rice oil used by the patients.   Second, the frying of foods with rice oil containing large amounts of PCB's, as practiced by the patients, could have given rise to new compounds in the oil, which might significantly alter the effective toxicity of the PCB's when ingested with them.   Third, the above-mentioned concentrations of Kanechlor 400 in the rice oil may not have been sufficiently accurate to allow a rigorous quantitative discussion, since the analytical methods for estimating PCB's in foods were not fully developed at that time.

2.8 Babies Born to Patients and to Non-affected Wives of Patients, and Babies Breast-fed by Patients

Thirteen women consisting of 11 with Yusho and 2 unaffected wives of patients, were shown to have delivered 10 live and 2 stillborn babies in Fukuoka Pref. between February 15 and December 31, 1968.[8-10]   As shown in Tables 2.8, 9 of the babies had unusually grayish or dark-brown skin, and similar pigmentation of the gingivae and nails was noted in 5 of them.   Increased discharge from the eyes was also notable in most cases. Histological examinations of a stillborn fetus showed marked hyperkeratosis and atrophy of the epidermis and cystic dilatation of the hair follicles, especially in the head region.   A marked increase in melanin pigment in the basal cells of the epidermis was also noted.[11]   A recent chemical analysis of this fetus has revealed that the concentrations of PCB's in the skin, mesenterial fatty tissue, and liver were not high, being 0.11, 0.02, and 0.07 ppm (on whole basis) and 1.2, 0.1, and 1.8 ppm (oy fat basis), respectively.[12]

Since no similar symptoms at birth, or at stillbirth, have been recorded in Japan, and their incidence was very high among the patients in question, these unusual phenomena were considered to be directly related to the mothers' ingestion of PCB-contaminated oil. However, a clear dose-response relationship between oil consumption and the phenomena could not be shown in these cases because of their limited number.   The amount of contaminated oil ingested by the mothers during pregnancy ranged from 0.3 to 2.6 l (Table 2.8).[8] Twelve of the 13 fetuses were smaller than the national standards, and 4 were small-for-date babies.[8,9]   As the living babies grew older, the skin pigmentation gradually faded.[10]   No evidence has so far been obtained with regard to any physical or mental handicap in them.   It is notable also that even 3 yr after the poisoning, babies borne by severe female patients tended to show darker stained skin on the back and gingivae, although the degree of darkness was less than that of earlier babies borne by the same mothers about 1 yr after the poisoning.[13]

Yoshimura has reported a particular case of a Yusho baby.   This child was born to a mother with moderate Yusho.   However, since the mother was considered to have ingested the contaminated rice oil only after delivery, i.e. when she was breast-feeding her baby, and the baby itself began to show symptoms of Yusho before ablactation (after 3–4 months of breast-feeding), the baby is thought to represent a very rare case of Yusho caused by the ingestion of PCB's exclusively through the mother's milk.[13] Nothing is known, however, concerning the possible concentration of PCB's in her milk due to the lack of any analysis.

TABLE 2.8
Details of Babies Born to Patients and Unaffected Wives of Patients, and Use of K Rice Oil by the Mothers

Mother No.	Age (yr)	Amount of oil used during pregnancy (1)	Pregnancy period when oil used	Grade of clinical severity of mother	Delivery	Sex of baby	Small-for-date	Stained skin	Stained gingivae	Stained nails	Increased eye discharge	Neonatal jaundice
1	28	0.3	3rd trimester	II	Normal	M	No	?	-	-	-	+
2	24	?	?	III	Normal	F	No	+	-	-	+	+
3	22	1.4	Later half of 2nd trimester	Normal	Forceps	M	No	?	-	-	+	++
4	27	0.7	2nd trimester	I	Normal	M	Yes	+	+	-	+	?
5	26	0.7	1st trimester	III	Normal	F	No	+	+	+	+	+
6	29	1.4	Later half of 2nd trimester	III	Normal	M	No	+	-	+	+	+
7	30	1.1	Throughout	I	Normal	M	Yes	+	-	-	+	?
8	23	?	Throughout	III	Normal	M	No	+	-	-	+	+
9	33	?	?	?	Caesarean section	M	Yes	+	+	+	+	?
10	29	0.3	Early half of 2nd trimester	Normal	Normal	F	No	?	-	-	-	+
11	26	?	?	II	Normal	M	No	+	+	+	+	+
12	33	2.6	Early half of 2nd trimester	IV	Stillbirth	?	?	+	?	?		
13	25	?	?	III	Stillbirth	F	Yes	+	?	?		

TABLE 2.9
PCB's in the Tissues of Yusho Patients

Patient	Date of death or surgical operation	Sex	Age	PCB's (ppm) on a whole basis (fat basis)						Cheese-like discharges of acneform eruptions	Ref.
				Skin	Fat tissue[†1]	Liver	Heart	Kidney	Brain		
A	Nov., 1968	—	adult							45.8[†2]	3
B	Nov., 1968	—	adult							32.1[†2]	3
C	Nov., 1968	M	18		s. c. face: 75.5 abdomen: 13.1						3
D	July, 1969	M	13		m. 1.3 (3.7)	0.14(9.5)	0.6(15.0)	0.1(9.6)	0.02(0.6)		12, 17
E	July, 1969	M	25	1.2(8.7)	m. 2.8(15.1)	0.2(10.4)	5.2(18.3)				12, 17
F	Nov., 1969	M	73	1.0(4.4)	m. 3.8(8.4)	0.07(3.1)					12, 17
G	Dec., 1970	F	48	0.6(0.8)	m. 0.7(0.9) g. o. 0.7(0.8)	0.07(1.3)	0.2(0.8)				12, 17
H	May, 1972	M	46	1.8(3.2)	m. 4.3(6.5)	0.08(8.4)	0.08(0.3)	0.01(0.4)			12, 17
I	Sept., 1972	F	33		s. c. 1.9(2.9) g. o. 1.0(2.2)						12, 17

†1 s. c.=subcutaneous; m.=mesentery; g. o.=great omentum.
†2 Calculated from the chlorine content.

2.9   GROWTH OF YUSHO CHILDREN

To determine whether Yusho had any disturbant effect on growth in children, 42 affected school-children (23 boys and 19 girls) were compared in 1967, 1968, and 1969 with their healthy classmates (719 children, matched by sex).   The gains of affected boys, in both height and weight, decreased significantly after poisoning, although affected girls showed no clear change in this respect.[14]

2.10   PCB's IN YUSHO PATIENTS

PCB concentrations in the tissues of Yusho patients were shown to be very high soon after the attack, as indicated by figures obtained by analyzing the subcutaneous fat tissue of the face and abdomen, and the cheese-like discharges of acneform eruptions (Tables 2.9).   Analysis of various tissue samples from patients who died, or who had surgical operations, also suggested that the early elevated concentration decreased fairly rapidly after discontinuation of use of the contaminated oil, although no accurate consecutive comparison was possible because of the lack of follow-up data for these patients.   It is worthy of note, however, that complete excretion of ingested PCB's takes an extremely long time.   This has been clearly indicated by the findings of Masuda et al.[15] and Takamatsu et al.,[16] who showed that the PCB concentration in the blood of Yusho patients did not return to a normal level even after the lapse of 5 yr from poisoning.

Another significant fact is that about 60% of tested patients showed a very unique, common gas-chromatographic pattern for PCB's in various tissues including the blood, which distinctly differed from the pattern of PCB's in ordinary persons.   Such a peculiar pattern could also be seen about 5 yr after the cessation of use of the contaminated oil.[12,15-17]

2.11   DEATHS AMONG YUSHO PATIENTS

Up to September 13, 1973, a total of 22 deaths had occurred among 1200 Yusho patients throughout Japan.   The causes of death, as reported by Urabe,[1] are shown in Table 2.10.   Malignant tumors accounted for 41% of the deaths, suggesting a possible excess mortality from them.   How-

TABLE 2.10
Causes of Death in Yusho Patients

Cause		No. of cases
Malignant neoplasms		9
Stomach cancer	*2*	
Stomach cancer + liver cancer	*1*	
Liver cancer + liver cirrhosis	*1†*	
Lung cancer	*1*	
Lung tumor	*1*	
Beast cancer	*1*	
Malignant lymphoma	*2*	
Cerebrovascular lesions		3
Amyloidosis		1†
Osteodystrophia fibrosa		1†
Myocardial degeneration + pericarditis		1†
Status thymicolymphaticus		1†
Liver cirrhosis		1
Suicide		1
Senility		1
Traffic accidents		3
Total		22

† Autopsied.

ever, no conclusive evidence for this is yet available, due to the lack of any rigorous statistical analysis.

REFERENCES

1. H. Urabe, *Fukuoka Acta Med.*, **65**, 1 (1974).
2. M. Kuratsune *et al.*, **60**, *ibid.*, 513 (1969).
3. M. Goto and K. Higuchi, *ibid.*, **60**, 409 (1969).
4. T. Yoshimura, *ibid.*, **62**, 104 (1971).
5. H. Tsukamoto *et al.*, *ibid.*, **60**, 496 (1969).
6. T. Kojima, *ibid.*, **62**, 25 (1971).
7. M. Kikuchi, Y. Mikagi, M. Hashimoto and T. Kojima, *ibid.*, **62**, 89 (1971).
8. A. Yamaguchi, T. Yoshimura and M. Kuratsune, *ibid.*, **62**, 117 (1971).
9. I. Taki, S. Hisanaga and Y. Amagase, *ibid.*, **60**, 471 (1969).
10. I. Funatsu *et al.*, *ibid.*, **62**, 139 (1971).
11. M. Kikuchi, M. Hashimoto, M. Hozumi, K. Koga, S. Oyoshi and M. Nagakawa, *ibid.*, **60**, 489 (1969).
12. Y. Masuda, R. Kagawa and M. Kuratsune, *ibid.*, **65**, 17 (1974).
13. T. Yoshimura, *ibid.*, **65**, 74 (1974).
14. T. Yoshimura, *ibid.*, **62**, 109 (1971).
15. Y. Masuda, R. Kagawa, K. Shimamura, M. Takada and M. Kuratsune, *ibid.*, **65**, 25 (1974).
16. M. Takamatsu, Y. Inoue and S. Abe, *ibid.*, **65**, 28 (1974).
17. Y. Masuda, R. Kagawa and M. Kuratsune, *Bull. Environ. Contam. Toxicol.*, **11**, 213 (1974).

3

Toxicological Aspects I: the Toxicity of PCB's

Kiichi UEDA

3.1 INTRODUCTION

Technical PCB's constitute a group of chlorine-substituted biphenyl homologs, and each commercial product is assigned a code number indicating the chlorine content (%) or number of substituent chlorine atoms of the predominant homolog. Besides this representative homolog, there are usually certain amounts of other PCB homologs with more or less chlorine atoms per molecule. Moreover, the content of highly toxic byproducts (such as polychlorinated dibenzofurans) coexistent in such technical products may vary according to the mode of synthesis. The results of animal experiments or effects on man may thus differ even in the case of equivalent types of PCB's of different brands. The discrepancies encountered among many previous reports are probably attributable in part to this chemical complexity.

The principal target organs of PCB's in man appear to be the liver and skin. However, in rodents, skin disorders are generally not apparent.   Toxicity studies in these animals have thus tended to concentrate on the liver, including morphological, biochemical, metabolic and electron-microscopic studies.   The tumorgenicity of PCB's for the liver of mice has also been investigated.   Other key problems in animals have included the effects on successive generations, such as reductions in pup number and teratological effects.   However, the mutagenicity of PCB's, i.e. their effects on chromosomes, has not yet been investigated in detail.

A number of good reviews has been published.   These would serve to give the reader a broad understanding of the health hazards of PCB's. The reviews include *"Polychlorinated biphenyls and the environment"*,[1] *"PCB's environmental impact"*,[2] the review by Kimbrough,[3] and *"Hazards to health and ecological effects of PCB's in the environment"* (WHO Regional Office for Europe).[4]

3.2   Acute Toxicity for Experimental Animals (Oral LD_{50})

The oral LD_{50} (50% lethal dose) of PCB's for albino rats has been reported to be several grams/kilogram body weight.   Highly chlorinated biphenyls such as tetrachlorobiphenyl and higher homologs are slightly less toxic (*ca.* 10 g/kg b.w.) than trichlorobiphenyl (5–6 g/kg b.w.).[5] Also, low chlorinated biphenyls are more soluble in water (even if only slightly) than the higher chlorinated homologs, so that their intestinal absorption may be accelerated.   Recently, Grant *et al.*[6] have reported the LD_{50} of Aroclor 1254 to be a few grams/kilogram body weight in Wistar strain rats, i.e. 1.4 g/kg b.w. for 30-day-old rats, and 2.0–2.5 g/kg b.w. for 120-day-old rats.   There were no major differences between the sexes.

The reason for the discrepancy in these two sets of LD_{50} data is not explained at present.   However, possible factors include the difference in rat strain used, the mode of administration, etc.

The major components of the PCB residues accumulating in the human body are penta- and hexachlorobiphenyls.   Their chronic toxicity is thus the key problem in evaluating possible health effects, and acute toxicity values are in this respect not of primary significance to work on the effects of the PCB's polluting the environment.

3.3   Subacute and Chronic Studies

3.3.1   Short-term, low-level exposure in nonhuman primates

Recently, Allen *et al.*[7] have fed PCB's (Aroclor 1248) at a level of 25 ppm for 2 months to 6 adult female rhesus monkeys.   The total dose of PCB's was between 250 and 400 mg for an average body weight of 5.6 kg. Facial edemas, alopecia and acne developed within 1 month, and one of the animals died from the effects of PCB intoxication 2 months after removal from the experimental diet. This animal had freely ingested about 300 g food/day during the exposure period, in comparison with about 175 g/day in the other 5 animals, and had then suffered from acute anorexia. The dead monkey showed symptoms of anemia, hypoproteinemia, severe hypertrophic-hyperplastic gastritis and bone-marrow hypoplasia. The thickness of the stomach wall was found to average 2 cm in the dead monkey, in comparison with a normal thickness of 2 mm, a change which could have induced the observed anorexia, loss of body weight and, ultimately, death. The surviving animals showed apparent acne, subcutaneous edemas and alopecia with 34 µg/g PCB in adipose tissues even 8 months after removal from the experimental diet, so indicating a persistence of toxic effects long after the termination of exposure.

3.3.2 Pathological changes in long-term feeding tests

As indicated above, the liver is the target organ of PCB's in rodents, and hypertrophy, fat degeneration and necrosis may result. A 2-yr feeding test in the Industrial Bio-test Laboratory in Chicago[8] revealed no effects on rat organs after feeding Aroclor 1242 at concentrations of 1, 10 and 100 ppm. Liver atrophy was observed on feeding Aroclor 1254 or 1260 at a concentration of 100 ppm, but there were no adverse histopathological effects. No changes were found at 1 or 10 ppm. The feeding of Aroclor 1254 at 100 ppm led to liver hypertrophy in both male and female beagle dogs, but at 10 ppm the effects were seen only in females. Hypertrophy itself is an adaptive reaction; however, in chickens, the most sensitive condition was found to be hydropericardium, so indicating different specific reactions according to the species. Reductions in organ weight in the kidney, liver and spleen were also found, and the morphological effects in birds in general could be said to be more pronounced than those in animals.

3.3.3 Electron-microscopic findings

Nishizumi et al.[9,10] have reported an increase in smooth-surfaced endoplasmic reticulum (SER), together with a decrease in rough-surfaced endoplasmic reticulum (RER), in liver cells of mice given Kanechlor 400 for 13 weeks (Fig. 3.1). Similar changes were also found in monkeys. Yamamoto[11] suggested that hypertrophy of the SER may be attributable in part to an increased synthesis of membrane proteins by ribosomes attached to the RER. Nishihara and Yamamoto[12] have also reported in

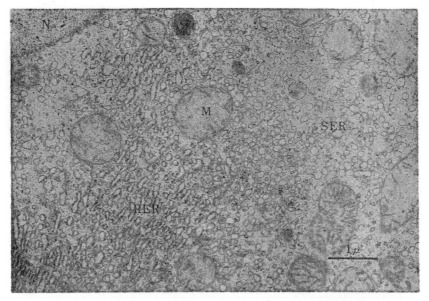

Fig. 3.1 Part of a hepatocyte from a mouse given 0.2 ml of 0.5% v/v Kanechlor 400 in olive oil for 13 weeks, showing an increased amount of SER. The RER was decreased, with reorganization into rough-surfaced sinuous cisternae. Mb, microbody; M, mitochondrion; N, nucleus. (Osmium fixation, uranyl acetate staining, ×51,000; photo provided by Dr. M. Nishizumi.)

guinea-pigs that SER proliferation was found even 2 months after a single dose of 100 mg/kg of PCB's (Kanechlor 400), whereas in rats the reticulum recovered its normal appearance 1 month after a single dose of 200 mg/kg. Vos and Notenboom-Ram[13] have recently reported proliferation of the SER which resulted in perinuclear and peripheral displacement of the mitochondria and RER. The focal cytoplasmic hyalin degeneration without glycogen observed under the light microscope was found by electron microscopy to be a densely packed agglomeration of SER, which appeared to represent hypertrophic, hypoactive SER.

3.4   Biochemical Reactions

3.4.1. Induction of drug-metabolizing enzymes in the liver

It is well known that chlorinated organic pesticides such as DDT, dieldrin, etc. are strong inducers of detoxicating enzymes in the liver. PCB's may also display a similar effect.

Street *et al.*[14] have studied the effects of feeding Aroclors (50 and 100 ppm) to female rats for 15 days.   The sleeping time induced by hexobarbital was shortened.   With Aroclor 1221 (50 ppm) there was an 11% reduction, and with Aroclors 1248 and 1268 there were corresponding reductions of 35 and 48%, respectively.   Aniline hydroxylation and *p*-nitroanisol demethylation were increased.   Also, the induction of microsomal hydroxylating enzymes has been reported by Lincer and Peakall[15] in the American kestrel, and by Risebrough *et al.*[16] in pigeons.   Villeneuve *et al.*[17] have reported that the no-effect level of Aroclor 1254 for enzyme induction in pregnant rabbits is 1.0–10 mg/kg when administered over a 28-day period during gestation.   Induction of aniline hydroxylase and aminopyrine *n*-demethylase was observed at 10 mg/kg Aroclor.

Fujita *et al.*[18] have compared the relative effects of di-, tetra-, penta- and hexachlorobiphenyls, and found that 2,4,5,3',4'-pentachlorobiphenyl was the most potent inducer among the homologs tested.   Komatsu and Tanaka[19] have studied the effects of small PCB doses on microsomal enzymes.   They found that in female rats a dose of 2 mg/kg/day of Kanechlor 500 for 3 consecutive days (total 6 mg/kg) led to a 49% reduction in the hexobarbital-induced sleeping time, as compared with the controls.   When the same total dose was given over a 15-day period (i.e. at 0.4 mg/kg/day), no reduction in sleeping time was observed, and when such administration was continued for 45 and 53 days the shortening did not exceed 12–13%.   (The dose of 0.4 mg/kg/day was estimated as equivalent to about 5 ppm feeding.)

Grant *et al.*[20] have fed Aroclor 1254 (0, 2, 20 and 100 ppm) to male rats for 8 months to study its effects.   The 20 ppm diet significantly decreased vitamin A storage and increased the hepatic aniline hydroxylase activity.   Such activity was also increased by the 2 ppm diet after 4 and 6 months.   Litterest *et al.*[21] have reported feeding tests on male rats with Aroclors 1242, 1248, 1254 and 1260 (0, 0.5, 5, 50 and 500 ppm).   The animals were sacrificed after 4 weeks of administration and 16 hours of fasting.   The largest increase in liver triglyceride was seen with Aroclor 1248, and the peak levels decreased with increasing chlorine content. The activity of liver microsomal demethylase was increased by all 4 Aroclors, at 5 and 50 ppm, and by Aroclors 1254 and 1260 at 0.5 ppm. There was a 2–3-fold increase at 500 ppm.   Nitroreductase activity was increased 50–70% by all 4 Aroclors, at 0.5 ppm, and Aroclor 1260 gave a 10-fold increase at 500 ppm.   Cytochrome P-450 activity was increased by Aroclors 1254 and 1260 at 5 ppm, and by all 4 Aroclors at 50 and 500 ppm.   Liver glucose-6-phosphatase was found to be inhibited by Aroclor feeding.   Benthe *et al.*[22] have reported the induction of oxidative demethylation in rats by single intraperitoneal injections of 5–500 mg/kg Aroclors 1232 and 1248.   The effects were still evident 4 weeks after the injection.

Recently, Villeneuve *et al.*[23] have described chronic PCB ingestion studies in rats.   They found that Aroclors 1254 and 1260 both consistently reduced the pentobarbital-induced sleeping time.   Vos *et al.*[24] have also reported an increase in cytochrome P-450 activity in the Japanese quail after administration of 100 mg/kg Aroclor 1260.

The induction of drug-metabolizing enzymes in the liver is an adaptive reaction of organisms for enhancing detoxication.   However, if the stimulus of foreign substances is very strong or of long duration, it may cause liver cell damage.   Moreover, if the induction occurs too strongly, even essential hormones (such as sex and steroid hormones) may be metabolized and so lose their physiological activity.

3.4.2   Hepatic porphyria

Vos and Koeman[25] have reported the occurrence of increased fecal excretion of coproporphyrin and protoporphyrin, and tissue fluorescence in chickens fed technical PCB's at 400 ppm.   Similar effects were also demonstrated in rabbits after dermal application of 118 mg PCB over a 38-day period (total 27 applications).   Kimbrough[26] has observed porphyria in rats fed 100 ppm of Aroclor 1254 or 1260.   Moreover, Vos *et al.*[24] have determined the relationship of δ-aminolevulinic acid (ALA) synthetase activity to hepatic porphyria in the Japanese quail after oral administration of Aroclor 1260.   The liver mitochondrial ALA synthetase activity was increased 10- or 20-fold at Aroclor levels of 50 or 500 mg/ kg, respectively.   Also, at 1 mg/kg, the activity increase above the control was statistically significant. The no-effect level was 0.1 mg/kg (mean liver PCB residue level <0.54 ppm) in this series of experiments.

3.4.3   Effect on lipid levels in the blood and other tissues

Stimulated by the finding of elevated serum lipid levels in Yusho patients, Japanese researchers have attempted to determine the effects of PCB's on lipid levels (particularly triglyceride levels in the plasma and skin lipids) in experimental animals.   Tanaka *et al.*[27] have found a 10-fold increase in the serum triglyceride level of rats after giving 0.1 g/kg Kanechlor 400 for 4 weeks.   Nagai *et al.*[28] fed rats orally with 0.5, 2.5, 5 and 50 mg Kanechlor 400 per animal for 2–6 weeks, and found that plasma triglyceride levels were elevated in the 50 mg/animal group.   Increases in plasma cholesterol and phospholipid levels were observed in the 0.5 mg/animal group, and these levels were further elevated at higher doses.   By comparing the low- and high-boiling fractions of PCB's, it was found that the latter fraction markedly reduced both the total amount of lipid present in abdominal fatty tissues and the plasma triglyceride levels. Ito *et al.*[29] have fed rabbits perorally with 8 ml of 1 % chlorobiphenyl olive oil solution for 3 or 11 days, and analyzed the subsequent lipid composition of the liver and plasma. They found an almost 2-fold increase in the total lipid and triglyceride levels, and, in the case of the 11-day administration, a remarkable increase in the serum triglyceride level.

Litterest *et al.*[21] have analyzed the liver triglyceride levels of rats after feeding them one of the Aroclor series (1242 through 1260) at concentrations of 0, 0.5, 5, 50 and 500 ppm for 4 weeks. The largest increases in liver triglyceride were found with Aroclor 1248, and the peak levels decreased with increasing chlorine content.

Nagai *et al.*[30] have administered Kanechlor 400 orally to rats at a level of 1 mg/animal (equiv. 6.7 mg/kg) for 30 days. Following an 18-hr fast, 1-^{14}C acetate was injected intraperitoneally in order to determine the rate of incorporation in lipids. Reductions in free fatty acid and triglyceride levels occurred in skin lipids, and the incorporation of acetate into triglyceride and free fatty acid was reduced. On the other hand, incorporation into sterols was increased, suggesting an activation of chain elongation and desaturation of fatty acids by PCB's.

Summarizing the above experimental results, it can be said that in animals a considerably high PCB dose is required for producing elevated plasma or serum triglyceride levels, indicating that it is not a very sensitive reaction. Reductions of triglyceride and free fatty acids were found in skin lipids, on the other hand. Further studies are required to interpret these apparently contradictory findings.

3.5 ENDOCRINE EFFECTS

Bitman and Cecil[31] have demonstrated an estrogenic activity for PCB's (Aroclor 1248) in the rat. They utilized the 18-hr glycogen response of the immature uterus after a single subcutaneous injection to evaluate the estrogenic activity, and found the minimum effective dose to be 8 mg. Also, higher chlorinated PCB mixtures were found to be inactive at this level. The possibility that the estrogenic effect of PCB's is due to *in vivo* conversion to hydroxylated analogs is suggested by the fact that *o,o'*-biphenol is a more potent estrogenic agent than chlorinated analogs (PCB's).

Komatsu[32] has reported no estrogenic activity for PCB's (0.1 g/kg Kanechlor 400, perorally for 3 days) in castrated rats. However, pretreatment with PCB's potentiated the action of estradiol to increase uterine weight in such castrated rats.

3.6   Immunosuppressive Activity

Based on the finding in experimentally poisoned animals of lymphoid tissue atrophy, lymphopenia, atrophy of the thymic cortex, and reductions in the number of germinal centers in the lymph nodes, Vos and de Roy[33] have investigated the immunosuppressive activity of PCB's. Guinea-pigs were fed Aroclor 1260 (0, 10 and 50 ppm) for 8 weeka, and half the animals in each group were given a tetanus toxoid injection in the right foot-pad to stimulate the lymphoid system.   Serum γ-globulin levels were significantly decreased in the toxoid-stimulated group fed 10 ppm PCB.   By a fluorescent antibody technique, γ-globulin-containing cells in the popliteal lymph nodes were shown to be reduced in this group, as they were also in the unstimulated lymph nodes (left side) of the same group and in such lymph nodes of toxoid-stimulated animals fed 50 ppm PCB. Vos and van Driel-Grootenhuis[34] also found decreased serum antitoxin titers in guinea-pigs treated with 50 ppm PCB when immunized with tetanus toxoid.   Depression of cellular immunity was demonstrated using the delayed-type hypersensitivity to tuberculin as a parameter.

Friend and Trainer[35] have fed Aroclor 1254 (0, 25, 50 and 100 ppm) to ducks for 10 days and then challenged them intraperitoneally with duck hepatitis virus ($1.5 LD_{50}$). Mortality was significantly higher in the PCB-treated birds than in the controls, but no differences were noted according to dose level.

Saito et al.[36] have reported related research on human patients of PCB poisoning (Yusho).   Serum IgA and IgM levels were decreased and IgG levels increased in 1970 (i.e. 2 yr after the outbreak), but were generally in the normal range by 1972.   The IgA level was remarkably reduced in a few cases showing respiratory symptoms (to below 50 mg/ 100 ml), suggesting a decreased resistance to respiratory infection.   The IgM level was significantly lower in patients suffering from severe dermatological symptoms.

3.7   Carcinogenesis

In 1973, Nagasaki, Tomii et al.[37] reported the successful induction of hepatomas in mice (7 of 12 animals, 58.3%) after feeding 500 ppm Kanechlor 500 for 32 weeks.   No tumors were observed at lower levels of Kanechlor 500, or with other Kanechlors (300 and 400), or in the controls (Table 3.1).

Nagasaki et al.[38] have demonstrated an additive effect for liver tumor

TABLE 3.1

PCB-Induced Tumors in Mice (From Nagasaki, Tomii et al.[37])

PCB	Concn. in diet (ppm)	No. of mice	Liver wt (g)	Nodular hyperplasia	Hepatoma (%)	Amyloidosis (%)
Kanechlor 500	500	12	5.06†	+	7/12 (58.3)	0/12
	250	12	3.16	−	0	2/12
	100	12	2.65	−	0	3/12
Kanechlor 400	500	12	3.01	−	0	0/12
	250	12	2.85	−	0	3/12
	100	12	2.80	−	0	10/12
Kanechlor 300	500	12	2.71	−	0	1/12
	250	12	2.11	−	0	4/12
	100	12	2.86	−	0	10/12
Control	−	6	1.45	−	0	0/6

† The liver had a rough surface with multiple tumors up to 0.2—1.0 cm in diameter. Some areas of nodules showed an adenomatous pattern, and many necrotic foci were present. Nuclear irregularities and mitotic figures were common in non-tumorous areas of the liver.

induction with simultaneously fed PCB's (Kanechlor) and α-BHC (see Table 3.2). Individual feeding of either chemical alone at the same levels did not lead to hepatoma formation.

Kimura and Baba[39] have reported benign tumors in female rats (only) after feeding a total of 1200–1500 mg Kanechlor 400 over a 400-day peri-

TABLE 3.2

Changes in the Liver of Male dd-Mice Treated with α-BHC and/or PCB's for 24 Weeks (From Nagasaki et al.[38])

BHC or PCB's in diet (ppm)	No. of mice	Liver wt. (% of body wt.)	Histology of liver			Liver nodules (%)	
			Oval cells	Bile duct prolif.	Cellular hypertrophy	Nodular hyperplasia	Hepatocellular carcinoma
α-BHC (250)	38	8.2±1.8	±	±	⧺	30 (78.9)	10 (26.3)
α-BHC (250) + PCB's-5 (250)	20	18.9±5.3	+	+	⧺	16 (80.0)	11 (55.0)
α-BHC (250) + PCB's-4 (250)	30	13.9±3.3	±	±	⧺	26 (86.7)	15 (50.0)
α-BHC (100)	20	5.3±0.5	−	−	+	0 —	0 —
α-BHC (100) + PCB's-5 (250)	25	10.6±2.0	±	±	⧺	8 (32.0)	1 (4.0)
α-BHC (100) + PCB's-5 (100)	24	7.6±1.3	−	−	+	3 (12.5)	0 —
α-BHC (100) + PCB's-4 (250)	29	7.6±1.1	−	−	+	4 (13.8)	0 —
α-BHC (100) + PCB's-4 (100)	27	6.1±0.6	−	−	+	0 —	0 —
α-BHC (50)	20	5.0±0.6	−	−	±	0 —	0 —
α-BHC (50) + PCB's-5 (250)	30	7.8±1.2	−	−	⧺	9 (30.0)	2 (6.7)
α-BHC (50) + PCB's-5 (100)	28	6.3±1.2	−	−	+	0 —	0 —
α-BHC (50) + PCB's-4 (250)	28	6.5±1.2	−	−	+	0 —	0 —
α-BHC (50) + PCB's-4 (100)	27	5.4±0.5	−	−	±	0 —	0 —
PCB's-5 (250)	20	7.5±1.1	±	±	+	0 —	0 —
PCB's-5 (100)	20	6.8±1.0	−	−	±	0 —	0 —
PCB's-4 (250)	20	7.5±1.6	±	±	+	0 —	0 —
PCB's-4 (100)	20	7.4±0.3	−	−	±	0 —	0 —
CONTROL	20	4.0±0.2	−	−	−	0 —	0 —

od.   The PCB dose was elevated step-wise commencing at 38.5 ppm, increasing to 462 ppm, and then maintaining this level for 32 weeks.   Pinhead- to pear-sized round and pale-brown flecks or nodules were found scattered on the liver surface (and also its cut surface) in animals ingesting more than a total of 1200 mg Kanechlor 400.   The growths were found microscopically to be adenomatous nodules, which were interpreted to be benign neoplastic lesions.

Recently, Allen and Norback[40] have fed male rhesus monkeys with a diet containing 300 ppm PCB (Aroclor 1248) or 5000 ppm PCT (polychlorotriphenyl) for 3 months.   This treatment led to hyperplasia and dysplasia of the gestric mucosa, and invasion of the adjacent tissue region, changes which are suggestive of an eventual neoplastic transformation.

3.8   EFFECTS ON SUCCESSIVE GENERATIONS

3.8.1   Reproduction tests

In a reproduction study sponsored by Monsanto (cited in ref. 1), Aroclors 1242, 1254 and 1260 were fed to rats at levels of 1, 10 and 100 ppm in their diet.   Aroclor 1242 showed no effects on the first generation, but the mating indices were decreased in the second generation at the 100 ppm level.   With Aroclor 1254, there were reductions in the number of pups delivered and the numbers of pups surviving at both the second and third litters.   Aroclor 1260 caused an increase in the number of stillbirths at the 100 ppm level.   No effects were observed at 1 or 10 ppm.

Ikeda[41] has reported results for reproduction tests in mice.   At 75 or 25 mg/kg/day of PCB, decreases in pup survival rate and average fetal body weight were observed, and 4 fetuses of abnormal form were noted. Moreover, a decreases in pup survival rate occurred even at a PCB dose of 3 mg/kg/day.

3.8.2   Placental transfer of PCB's

Placental transfer of PCB's into the developing fetus has been demonstrated by Grant et al.[42] in pregnant rabbits after feeding Aroclor 1221 and 1254.   The fetal liver showed higher PCB residue levels than the maternal liver.

In their experiments on rhesus monkeys, Allen et al.[7] have reported the case of an infant born to a mother 3 months after termination of PCB feeding to her.   The liver and fatty tissues of the mother contained about 50 ppm PCB's, whereas the levels in the liver and fat of the infant were 0.01 and 27.7 ppm, respectively.   The adrenals of the infant showed a high PCB level (24.4 ppm), but the levels in other organs were less than 1 ppm.   Although this information is for only a single animal, it is significant as an experiment on a primate species.

3.8.3   Teratology

Teratogenicity studies have been conducted by Keplinger *et al.*[43] in rats, using Aroclor 1242, 1254 and 1260.   However, daily oral doses of 10 and 30 mg/kg on days 6 through 15 of the gestation period yielded no observable teratogenic effects.   On the other hand, McLaughlin *et al.*[44] have noted teratogenic effects after injecting 5 mg Aroclor 1242 into the yolk sac of chicken eggs.

Villeneuve *et al.*[45] have reported that oral administration of Aroclor 1254 to rabbits at 12.5 mg/kg/day for the first 28 days of pregnancy caused abortions.   However, no consistent skeletal abnormalities were found in the dead fetuses.   Aroclor 1254 has been given to pregnant rats at levels of 0.1 and 50 mg/kg (once per day from the 7th to 15th day of organogenesis) by Curley *et al*[46]   No significant effects were found on total litter weights, stillbirth rate, or the survival rate to weaning.

3.8.4   Mutagenicity

Keplinger *et al.*[43] have reported that a dominant lethal assay in rats gave no evidence of mutagenic effects for PCB's.   Also, Hoopingarner *et al.*[47] have observed no chromosomal aberrations in human lymphocyte cultures exposed to 100 ppm Aroclor 1254.

On the other hand, Peakall *et al.*[48] have reported that embryos from the second generation of ring doves fed 10 ppm Aroclor 1254 exhibited a high frequency of chromosomal aberrations.   Such a level of dietary PCB's was also associated with a high frequency of embryonic deaths.

3.9   TOXICOLOGICAL SIGNIFICANCE OF COEXISTENT BYPRODUCTS OF HIGHER TOXICITY

In 1970–71, Vos *et al.*[25,49] reported a remarkable difference in toxicity between different brands of technical PCB's in chickens and rabbits.   Clophen A60 and Phenoclor DP60 caused death, centrolobular liver necrosis and subcutaneous and abdominal edemas in chickens, whereas such findings were only occasionally observed with Aroclor 1260.   Vos *et al.* attributed this difference to the coexistence of other more toxic substances, and the presence of tetra- and pentachloro dibenzofurans in Clophen and Phenoclor was finally established.

Recently, certain chemically pure PCB isomers have become available. Vos and Notenboom-Ram[13] have compared the toxicity of 2,4,5, 2',4',5'-hexachlorobiphenyl and Aroclor 1260 by applying them to the ears of rabbits (120 mg; 20 applications over 4 weeks). No differences were noted in the hepatotoxic action of the two substances, but the major acnegenic properties of the technical grade PCB appeared to arise from chlorinated dibenzofurans. The pure hexachlorobiphenyl was more porphyrogenic then the technical mixture. It is thus probable that hepatic porphyria arises from the effects of PCB's. In other experiments, Komatsu and Kikuchi[50] applied pure crystalline 3,4,3',4'-tetrachlorobiphenyl to the inner surface of the ears of rabbits. The resulting hyperkeratosis, dilatation of the hair follicles and cyst formation were identical to those induced by the commercial PCB mixture, Kanechlor 400, although the grade of reaction was stronger with the pure PCB.

Table 3.3 gives a generalized summary of the probable relative contributions of PCB's and chlorinated dibenzofurans in producing observable symptoms of poisoning, as presented by Vos and Notenboom-Ram.[13]

TABLE 3.3

Probable Contributions of Chlorinated Dibenzofuran and Pure PCB in the Toxicity of Technical PCB Preparations (From Vos and Notenboom-Ram[13])

	Chlorance	Edema formation	Liver damage	Hepatic porphyria
Chlorinated dibenzofuran	++	++	++	−
Polychlorinated biphenyl	+	−	+	++

3.10 COMPARATIVE TOXICITY OF SYNTHETIC ISOMERS

Recently, many chemically pure PCB isomers have been supplied as standard substances for gas-chromatographic analysis. They are of course very expensive, and comparatively large quantities are required at appropriate prices for the purposes of toxicological studies. However, such studies with synthetic isomers are gradually increasing.

Kawanishi, Sano and Mizutani[51] have compared 6 tetrachlorobiphenyl isomers and 3 hexachlorobiphenyl isomers by feeding them to male mice at 300 ppm for 14 weeks. After fasting for 1 day to enhance porphyrin production, the animals were sacrificed. Marked liver enlargement, PCB and protoporphyrin accumulation, and clear histopathological findings in the liver were observed in the case of 3,5,3',5'-tetrachlorobiphenyl and 2,4,6,2',4',6'- and 3,4,5,3',4',5'-hexachlorobiphenyls (see Table 3.4; cf. Fig. 3.2). Sano et al. proposed the hypothesis that chlorobiphenyls having 2 or more adjacent non-substituted carbon atoms on their rings are readily excreted ("non-accumulating"-type isomers, such as I, III, IV and VI in Fig. 3.2). However, if the 4-position is substituted

TABLE 3.4

Relative Effects[†1] of Various PCB Isomers[†2] and Commercial PCB's Fed to Mice (From Kawanishi et al.[54])

PCB	No. of Cl atoms	No. of deaths	Liver wt. (% of body wt.)	PCB level (μg/g)	Content of porphyrins (μg/g)		Light-microscopic findings		
					Proto-	Copro-	Fatty deg.	Pigment	Necrosis
I	4	0	5.92±0.27	<1	0.38±0.19	0.20±0.15	-	+	+
II	4	0	5.53±0.84	2.6±2.3	0.40±0.01	0.24±0.08	-	±	+
III	4	0	5.17±0.23	<1	0.48±0.34	0.13±0.09	±	-	-
IV	4	0	6.29±0.47	<1	1.00±0.78	0.17±0.15	-	±	±
V	4	4	8.99±1.67	2500±1600	0.88±0.52	0.13±0.07	±	-	±
VI	4	0	4.92±0.44	<1	0.25±0.17	0.06±0.05	-	±	-
VII	6	0	5.85±0.74	12±8	0.34±0.06	0.26±0.10	±	-	+
VIII	6	4	9.56±0.46	1800±560	2.00±0.91	0.37±0.31	+	-	+
IX	6	4	9.45±0.13	650±370	2.20±1.79	0.20±0.18	±	-	±
Kanechlor 400	4	1	6.82±0.30	86±46	0.57±0.22	0.33±0.02	+	+	+
Kanechlor 500	5	2	7.46±1.26	480±190	0.83±0.29	0.83±0.52	+	+	+
Kanechlor 600	6	2	9.48±0.70	3200±1600	1.29±0.14	0.68±0.08	±	±	+
CONTROL		0	4.87±0.55	<1	0.47±0.28	0.15±0.08	-	-	-

†1 Data represent means ± S.D. for groups of 4 mice each.   †2 See Fig. 3.2.

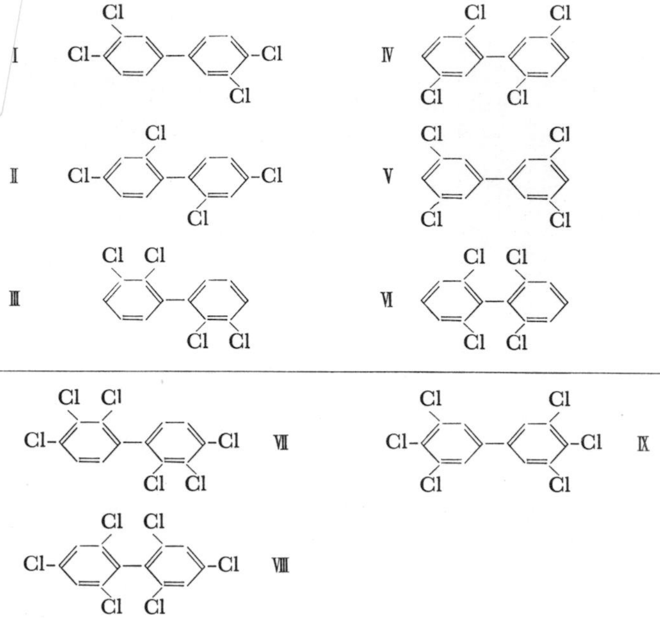

Fig. 3.2 Synthetic tetra- and hexachlorobiphenyl isomers.

with a chlorine atom, metabolism will be retarded due to steric disturbance ("intermediate"-type isomers, such as II and VII). Chlorobiphenyls having no adjacent pair of non-substituted carbon atoms on their rings represent "accumulation"-type isomers (V, VIII and IX).

3.11 POSSIBLE HEALTH EFFECTS : PRESENT AND FUTURE

As described in section 3, PCB's and various chlorinated organic pesticides are known to induce drug-metabolizing enzymes in the liver, and potent inducers such as DDT, dieldrin heptachlor, etc. are known to act additively or potentiatingly with PCB's. Present levels of these compounds in human adipose tissues vary widely from locality to locality, and at worst surpass the 20 ppm level. The possibility exists therefore that such levels could lead to abnormal enzyme induction in the liver, resulting in accelerated breakdown of essential hormones and other important biological substances. Indeed, the reduction in population of certain fish-eating birds has been attributed to this cause. However, in man, it has yet to be established that such effects could result from present environmental contamination.

Another possible effect of environmental PCB's is liver damage, which might occur via a mechanism similar to that with other organochlorine compounds. Also, PCB (pentachlorobiphenyl) is known to induce hepatomas in mice, as are α-BHC and p,p'-DDT; however, the experimental doses in these cases were high. Although it is not yet clear just how dangerous current pollution levels are in these respects, the known additive effects of chlorinated hydrocarbons in experimental carcinogenesis clearly indicate that monitoring of the total burden of such chemicals in human tissues is an urgent measure for protecting health.

Elevation of serum lipid (triglyceride) levels does not appear to be a very sensitive reaction in animals, although possible disturbance of lipid metabolism after intake of PCB's could have a relation to human atherosclerosis. The potential effects of low levels of PCB's on immunoreactions and other physiological functions in adult man remain uncertain.

Mother's milk has been shown to contain more than 0.06 ppm PCB's in certain parts of Japan, the U.S.A. and W. Germany. Considering the generally high sensitivity of babies to poisons, and the relatively high intake per kilogram body weight, infants may thus be most exposed to the possible risks, even at present levels of environmental PCB pollution. Fortunately, however, in 1974, the level in mother's milk was shown to have fallen to about 0.04 ppm even in the most polluted areas of Japan around the Seto Inland Sea, although it is uncertain whether the levels in local fish actually decreased or whether the mothers just controlled their intake of such fish. Be that as it may, the situation does appear to be improving in Japan due to the strict control measures imposed regarding the use of PCB's.

REFERENCES

1. Interdepartmental Task Force on PCB's, Washington D.C., *Polychlorinated Biphenyls and the Environment*, 1972.
2. *PCB's Environmental Impact, Environ. Res.*, 5 (no. 3), 1972.
3. R. D. Kimbrough, *Arch. Environ. Health*, **25**, 125 (1972).
4. WHO Regional Office for Europe, *The Hazards to Health and Ecological Effects of Persistent Substances in the Environment-Polychlorinated Biphenyls*, Copenhagen, 1975; National Institute of Environmental Health Sciences, *Environmental Health Perspectives*, Experimental Issue No. 1, pp. 1–185, April, 1972.
5. Data of Monsanto; cited in the FDA Status Report on the Chemistry and Toxicology of Polychlorinated Biphenyls (PCB) or Aroclors, 1970.
6. D. L. Grant and W. E. J. Phillips, *Bull. Environ. Contam. Toxicol.*, **12** (2), 145 (1974).
7. J. R. Allen, L. A. Carstens and D. A. Barsotti, *Toxicol. Appl. Pharmacol.*, **30**, 440 (1974).
8. Industrial Bio-test Laboratory, Chicago; cited as a Monsanto report in ref. 1.
9. M. Nishizumi, S. Kohchi and M. Kuratsune, *Fukuoka Acta Med.*, **60**, 539 (1969).
10. M. Nishizumi, *Arch. Environ. Health*, **21**, 620 (1970).
11. T. Yamamoto, *Cell Biol. Symp.*, **21**, p. 87, 1970.
12. H. Nishihara and T. Yamamoto, *Fukuoka Acta Med.*, **63**, 352 (1972).
13. J. G. Vos and E. Notenboom-Ram, *Toxicol. Appl. Pharmacol.*, **23**, 563 (1972).

14. J. C. Street, F. M. Urry, D. J. Wagstaff and A. D. Blau, *Abstr. Papers Am. Chem. Soc.*, no. 158, 1969.
15. J. L. Lincer and D. B. Peakall, *Nature*, **228**, 783 (1970).
16. R. W. Risebrough, P. Rieche, D. B. Peakall, S. G. Herman and M. N. Kirven, *ibid.*, **220**, 1098 (1968).
17. D. C. Villeneuve, D. G. Grant, W. E. J. Phillips, M. L. Clark and D. G. Clegg, *Bull. Environ. Contam. Toxicol.*, **6**, 120 (1971).
18. S. Fujita, M. Tsuji, K. Kato, S. Saeki and H. Tsukamoto, *Fukuoka Acta Med.*, **62**, 30 (1971).
19. F. Komatsu and K. Tanaka, *ibid.*, **62**, 35 (1971).
20. D. L. Grant, C. A. Moodie and W. E. J. Phillips, *Abstr. Papers Am. Chem. Soc.*, no. 164, 1972.
21. C. L. Litterest, T. M. Farber, A. M. Baker and E. J. Van Loon, *Toxicol. Appl. Pharmacol.*, **23**, 112 (1972).
22. H. F. Benthe, A. Schmoldt and H. Schmidt, *Arch. Toxikol.*, **29**, 97 (1972).
23. D. C. Villeneuve, D. G. Grant and W. E. J. Phillips, *Bull. Environ. Contam. Toxicol.*, **7**, 264 (1972).
24. J. G. Vos and R. B. Beems, *Toxicol. Appl. Pharmacol.*, **19**, 617 (1971).
25. J. G. Vos and J. H. Koeman, *ibid.*, **17**, 656 (1970).
26. R. D. Kimbrough, R. E. Linder and T. B. Gaines, *Arch. Environ. Health*, **25**, 354 (1972).
27. K. Tanaka, S. Fujita, F. Komatsu and N. Tamura, *Fukuoka Acta Med.*, **60**, 544 (1969).
28. J. Nagai, M. Furukawa, Y. Yae and K. Higuchi, *ibid.*, **62**, 42 (1971).
29. Y. Ito, H. Uzawa and A. Notomi, *ibid.*, **62**, 48 (1972).
30. J. Nagai, K. Higuchi and Y. Yae, *idid.*, **63**, 367 (1972).
31. J. Bitman and H. C. Cecil, *J. Agri. Food Chem.*, **18**, 1108 (1970).
32. F. Komatsu, *Fukuoka Acta Med.*, **63**, 374 (1972).
33. J. G. Vos and Th. de Roij. *Toxicol. Appl. Pharmacol.*, **21**, 549 (1972).
34. J. G. Vos, L. van Driel-Grootenhuis, *Sci. Total Environ. (Amsterdam)*, **1**, 289 (1972) cited in Ref. 4.
35. M. Friend and D. O. Trainer, *Science*, **170**, 1314 (1970).
36. R. Saito, N. Shigematsu and S. Ishimaru, *Fukuoka Acta Med.*, **63**, 408 (1972).
37. H. Nagasaki, S. Tomii, T. Mega, M. Marugami and N. Ito, *Gann*, **63**, 805 (1972).
38. H. Nagasaki, S. Tomii and T. Mega, *Nihon Eiseigaku Zasshi* (Japanese), **30** (1), 58 (1975).
39. H. Kimura and T. Baba, *Gann*, **64**, 105 (1973).
40. J. R. Allen and D. H. Norback, *Science*, **179**, 498 (1973).
41. Y. Ikeda, *Shokuhin Eiseigaku Zasshi* (J. Food Hygienic Soc. Japan), **13**, 359 (1972).
42. D. L. Grant, W. E. J. Phillips and D. C. Villeneuve, *Bull. Environ. Contam. Toxicol.*, **6**, 102 (1971).
43. M. L. Keplinger, O. E. Fancher and J. C. Calendra, *Toxicol. Appl. Pharmacol.* **19**, 402 (1971).
44. J. McLaughlin, G. P. Marliac, M. J. Kerrett, M. K. Murchler and O. C. Fitzhugh, *Toxicol. Appl. Pharmacol.*, **5**, 760 (1963).
45. D. C. Villeneuve, D. G. Grant, L. Khera, D. J. Clegg, H. Baer and W. E. J. Phillips, *Environ. Physiol.*, **1**, 67 (1971).
46. A. Curley, V. M. Burse and M. E. Grim, *Food Cosmet. Toxicol.*, **11**, 471 (1973).
47. R. Hoopingarner, A. Samuel and D. Krane, *Environ. Health Perspect.*, **1**, 155 (1972).
48. D. B. Peakall, J. L. Lincer and S. E. Bloom, *ibid.*, **1**, 1, 103 (1972).
49. J. G. Vos, J. H. Koeman, H. L. van der Maas, M. C. ten Noever de Brauw and R. H. de Vos, *Food Cosmet. Toxicol.*, **8**, 625 (1970).
50. F. Komatsu and M. Kikuchi, *Fukuoka Acta Med.*, **63**, 384 (1972).
51. M. Kawanishi, S. Sano and T. Mizutani, *Nihon Eiseigaku Zasshi* (Japanese), **30**, 124 (1975).

4

Toxicological Aspects II: the Metabolic Fate of PCB's and Their Toxicological Evaluation

Hidetoshi YOSHIMURA
and Shin'ichi YOSHIHARA

4.1 INTRODUCTION

Polychlorinated biphenyls (PCB's) have a wide range of industrial applications and are widespread in the environment. Since they exhibit a high degree of chemical and biological stability, and also lipid solubility, they tend to accumulate in food chains and have been detected in human tissues.[1] Further, as described in other chapters, a serious outbreak of PCB poisoning (Yusho) was recently experienced in southwest Japan. It was caused by the ingestion of specific lots of Kanemi rice oil which had accidentally been contaminated with large amounts of Kanechlor 400 (a commercial preparation of PCB's with a chlorine content of about 48%) during the manufacture of the oil.[2]   Before this occurrence of PCB intoxication, other evidence for the toxicity of PCB's was already known in the fields of ecology,[3,4] poultry science[5] and occupational medicine.[6,7]

Besides possibly threatening human health through direct contamination, PCB's are also considered a long-term potential hazard to the environment.   However, we do not as yet have sufficient knowledge regarding the absorption, distribution, metabolism and elimination of these substances.   In this chapter, therefore, a review will be given of the behavior of PCB's in the body of animals, in the hope that such biological findings may provide a clue towards gaining a fuller understanding of the mechanism of PCB toxicity in man.

4.2   ABSORPTION, DISTRIBUTION AND ELIMINATION OF PCB's

4.2.1   Absorption of PCB's from the gastrointestinal tract

It is widely recognized that PCB's can easily penetrate into biological systems and have a tendency to accumulate there, since they have been widely detected in the tissues of fish, wildlife, and even man.   Indeed, it has been shown in research on Yusho that considerable amounts of PCB components still remain in the adipose tissue, sebum and placenta of patients, and also in the fetuses of affected mothers, several months after becoming ill.[2]   These findings strongly suggest that PCB's are absorbed from the gastrointestinal tract, distributed to all tissues, and barely eliminated from the body during this interval.

At present, no conclusive evidence is available for the exact mechanism and rate of absorption of PCB's from the gastrointestinal tract, although it is considered that they may well be absorbed by a passive diffusion mechanism due to their lipid-soluble and neutral (non-ionizable) properties.   In an attempt to confirm this, the absorption of a PCB mixture from the gastrointestinal tract was studied *in vivo* in rats using ^{3}H-labeled Kanechlor 400,[8] together with the distribution in and elimination of PCB's from the tissues.[9]   Contrary to expectations, the results showed that such absorption was itself insufficient at the dose level used (25 mg/animal), the amount of absorbed PCB's being calculated at about half the dosage.   As shown in Table 4.1, most of the fecal radioactivity was eliminated during the first 2 days, although further continuous elimination of a small amount of radioactivity was observed over a long period.   Additional studies suggested that this radioactivity consisted largely of that from unchanged PCB's, with only a little from metabolites.

It is generally accepted that foreign compounds passing into the feces

TABLE 4.1

Excretion of Radioactivity into the Urine and Feces of Rats after Oral
Administration of ^{3}H-Kanechlor 400† (From Yoshimura et al.[9])

Sample	Days after ingestion	Radioactivity	
		(dpm)	% of dose
Urine	1	1247×10^4	1.12
	2	425	0.38
	3	173	0.16
	4—7	279	0.25
	8—14	97	0.09
	15—21	53	0.05
	22—28	7	0.01
	Total		2.06
Feces	1	35440×10^4	34.93
	2	17760	16.00
	3	4708	4.24
	4—7	10020	9.02
	8—14	2952	2.66
	Total		66.85

† ^{3}H-Kanechlor 400 (specific act. 20 μCi/mg) was orally administered to male
rats of Wistar King strain (wt. ca. 150 g) in a single dose of 25 mg/animal.

after oral administration are derived in two ways.   One is the direct passage of unabsorbed material from the gastrointestinal tract, and the other
is the passage of once-absorbed material into the feces via the biliary
system.   The latter route, however, may be ruled out in the present case
since, as described in a later section, it has been established using 2,4,3',4'-
tetrachlorobiphenyl that PCB's are not excreted unchanged into the bile.
From these facts, the radioactivity due to unchanged PCB's in the feces
must correspond mostly to that of unabsorbed material.

4.2.2   Distribution of PCB's in animal tissues

In the above experiment where ^{3}H-Kanechlor was administered orally to rats, it was also found that at 3 days after ingestion, maximum radioactivity was observed in the adipose tissue and skin, followed by the liver,
adrenals and gastrointestinal tract.[9]   All these tissues exhibited higher
radioactivity levels than the plasma.   After 4 weeks, most of the radioactivity had been eliminated from the tissues, although significant levels
were still detectable in some tissues even after 8 weeks (Table 4.2).

However, since commercial PCB preparations are complex mixtures
of biphenyls substituted at different positions and with different numbers
of chlorine atoms, the important question remained as to whether or not
these compounds all behaved in the same manner within the body.   To
answer this question, Yoshimura et al. traced the temporal changes of gas-
chromatographic patterns of residual PCB components in the tissues of

Table 4.2

Distribution of Radioactivity in Various Rat Tissues after Oral
Administration of ^{3}H-Kanechlor 400[†1] (From Yoshimura *et al.*[9])

Tissue	Days after ingestion	Radioactivity/100mg tissue (dpm)	Total radioactivity (dpm)
Brain	3	346×10^2	65×10^4
	28	16	2.7
	56	12	2.1
Lung	3	661	49
	28	47	6.7
	56	25	3.0
Liver	3	1230	908
	28	—	—
	56	—	—
Kidney	3	590	77
	28	—	—
	56	—	—
Adrenals	3	1035	4.0
	28	71	0.3
	56	33	0.2
Testis	3	453	99
	28	11	2.7
	56	6	1.6
Gastrointestinal tract	3	1000	1100
	28	5	8
	56	13	19
Heart	3	490	26
	28	17	1.3
	56	12	1.0
Spleen	3	450	17
	28	27	0.8
	56	23	1.1
Skin[†2]	3	2758	5840
	28	190	711
	56	112	527
Adipose tissue[†2]	3	2973	1141
	28	560	319
	56	821	591
Muscle	3	515	—
	28	11	—
	56	5	—
Bone	3	20	—
	28	1	—
	56	1	—

[†1] Experimental conditions were as described in Table 4.1.
[†2] The weights of total skin and adipose tissues were proposed to be 1/6 and 1/40 of the body weight, respectively.

TABLE 4.2—*Continued*

Tissue	Days after ingestion	Radioactivity/100mg tissue (dpm)	Total radioactivity (dpm)
Plasma[3]	3	842	314
	28	20	11
	56	8	5.3
Red blood cells[3]	3	47	9.7
	28	—	—
	56	—	—

[3] Plasma and red blood cell volumes were calculated using the following equations:
Plasma volume = body weight $\times 0.04 \times$ (1-hematocrit/100).
Red blood cell volume = body weight $\times 0.04 \times$ hematocrit/100.

mice for 10 weeks after the oral administration of Kanechlor 400.[10]   The chromatograms for kidney extracts at 1 day and 2 weeks after ingestion are shown together with that of the original Kanechlor 400 in Fig. 4.1.   As described in detail later, the peaks denoted *g–k*, and *l* and *m* were found to correspond to tetra- and pentachlorobiphenyls, respectively.[11]   The peak pattern at 1 day after ingestion clearly reflected that of the original Kanechlor 400; however, at 2 weeks, there was a significant difference. The peaks of shorter retention time (*a–g*) disappeared more rapidly than those of longer retention time (*h–n*).   Similar results were also obtained for other tissues.

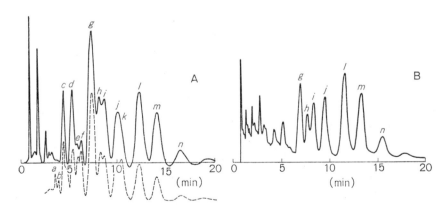

Fig. 4.1   Gas chromatograms of Kanechlor 400 (----) and kidney extracts of mice (——) killed 24 hr (A) and 2 weeks (B) after oral administration of Kanechlor 400 (2 mg/animal).   (From Yoshimura *et al.*[10]) Gas chromatography was carried out using an electron capture detector: 1.5% SE-30 on Chromosorb W (ϕ4 mm $\times$ 2m, glass column); column temp., 200°C; carrier gas, nitrogen at 22.7 ml/min flow rate.

4.2.3   Elimination of PCB's from the body of animals

The above findings suggested that the components of Kanechlor 400 were absorbed almost equally from the gastrointestinal tract and distributed rapidly in the same overall proportions in the tissues, but that they then disappeared at different rates from the tissues.   As shown in Table 4.3, tetrachlorobiphenyls (g and j), the principal components of Kanechlor 400, were found to be almost completely eliminated from the tissues within 3–4 weeks after oral administration at a dose of 2 mg per mouse, but penta- (l and m) and hexachlorobiphenyls (n) were still retained in small amounts after 9–10 weeks.[10]

Similar observations have been made by Grant et al.,[12] who found that Aroclor 1254 components with shorter retention times were eliminat-

TABLE 4.3

Elimination of the Main Components of Kanechlor 400 from the Tissues of Mice after Oral Administration (2 mg/animal) (From Yoshimura et al.[10])

Days after ingestion	Relative amounts of components[1,2]								
	(g)			(j)			(l)		
	Skin	Liver	Kidney	Skin	Liver	Kidney	Skin	Liver	Kidney
1	100	100	100	100	100	100	100	100	100
7	42	1	14	28	—	8	54	3	15
14	13	4	5	12	4	7	26	8	13
21	6	3	2	5	14	—	15	11	10
28	1	1	—	1	—	—	3	6	6
35	1	1	—	2	2	—	15	7	8
42	1	—	—	1	—	—	12	5	2
49	1	—	—	1	—	—	14	2	6
56	1	—	—	1	—	—	15	6	2
63	1	—	—	2	—	—	22	8	4
70	—	—	—	—	—	—	5	3	—

Days after ingestion	(m)			(n)		
	Skin	Liver	Kidney	Skin	Liver	Kidney
1	100	100	100	100	100	100
7	57	4	19	51	—	18
14	25	10	16	39	2	15
21	15	15	16	35	18	17
28	—	—	8	8	—	8
35	20	10	12	36	13	14
42	16	9	6	28	13	5
49	18	5	12	29	6	13
56	23	10	5	46	33	5
63	14	11	9	38	25	9
70	9	8	—	16	23	4

[1] The amounts of the residues in each tissue are expressed as values relative to those at 24 hr after ingestion ($=100$).

[2] The components represented as g, j, l, m and n corresponded to the gas-chromatographic peaks as labeled in Fig. 4.1.

ed to a greater extent than those with longer retention times, and that rats with a carbon tetrachloride-damaged liver were unable to eliminate the mixture as rapidly as rats with a normal liver. These findings strongly suggest that metabolism of PCB's proceeds mainly in the liver, and that this plays an important role in PCB elimination. The results are also in good agreement with the fact that the residual PCB's detected in contaminated animals are mainly penta- and hexachlorobiphenyl isomers.

As mentioned above, small amounts of radioactivity were excreted into the feces of rats for a long time after the oral administration of ^{3}H-Kanechlor 400. It seems unlikely that part of the original, orally administered compound which remained unabsorbed from the gastrointestinal tract, continued to be eliminated gradually over such a long period. One explanation might therefore be excretion through the biliary system; however, as mentioned, this possibility has been ruled out (see below). An alternative possibility is that unchanged PCB's might be excreted through the gut wall. In this connection, Williams et al.[13] have found that dieldrin injected intravenously into rats with the bile duct cannulated is excreted in part in an unchanged state into the feces, suggesting excretion of this organic chlorocompound from the gut wall. Shore et al.[14] have also established that basic compounds with a pK_a of 5.0 or more (e.g. aminopyrine) are excreted into the stomach, and other investigators[15,16] have suggested the excretion of certain other compounds into the gastrointestinal tract. These results together imply that PCB's may also be partially excreted in an unchanged state into the gastrointestinal tract.

This conclusion is supported by other findings, as follows: (1) Rats which were given 2,4,3',4'- or 3,4,3',4'-tetrachlorobiphenyl orally, eliminated in the feces small amounts of the unchanged compounds every day, together with their metabolites, for a period of more than 10 days[17-19] (Tables 4.4 and 4.5). (2) In the case of mice administered 2,4,3',4'-tetra-

TABLE 4.4
Fecal Excretion of Unchanged 2,4,3',4'-Tetrachlorobiphenyl and Its Major Metabolite (M—A_2) in Rats after Oral Administration (25 mg/animal) (From Yoshimura et al.[17])

Days after administration	Excretion rate (% of dose)†	
	Unchanged	M—A_2
1	37.8	0.9
2	1.8	2.5
3	0.4	1.7
4	0.4	1.3
5—6	0.8	0.7
7—8	0.8	1.7
9—12	1.3	1.8
Total	43.3	10.6

† Values are means of 3 rats.

TABLE 4.5

Fecal Excretion of Unchanged 3,4,3',4'-Tetrachlorobiphenyl and Its Major Metabolite (M—2) in Rats after Oral Administration (25 mg/animal) (From Yoshimura et al.[19])

Days after administration	Excretion rate (% of dose)†	
	Unchanged	M—2
1	58.6	1.6
2	4.9	0.9
3	0.4	0.4
4	0.1	0.2
5	0.1	0.1
6	Trace	0.1
7	Trace	Trace
8—14	Trace	Trace
Total	64.1	3.3

† Values are means of 3 rats.   Trace = <0.04.

TABLE 4.6

Fecal Excretion of Unchanged 2,4,3',4'-Tetrachlorobiphenyl and Its Major Metabolite (M—A$_2$) in Mice after Intraperitoneal Injection (250 mg/kg) (From Yoshimura et. al[20])

Days after administration	Excretion rate (% of dose)	
	Unchanged	M—A$_2$
1	1.1	0.3
2	1.3	0.3
3	1.6	0.6
4	1.3	0.7
5	1.3	0.7
6	1.1	0.7
7	1.1	0.6
Total	8.8	3.9

TABLE 4.7

Fecal Excretion of Unchanged 2,4,3',4'-Tetrachlorobiphenyl in Rats after Intravenous Injection (1.0 mg/animal)† (From Yoshimura et al.[22])

		(% of dose)		
0—24 hr	24—48 hr	48—72 hr	72—96 hr	Total
0.8	0.5	0.5	0.7	2.5

† Values are means of 3 rats.

chlorobiphenyl intraperitoneally, both the unchanged compound and its metabolites were excreted into the feces[20] (Table 4.6).   (3) Individual mono-, di-, tetra- and hexachlorobiphenyl isomers injected intraperitoneally into rats were in part excreted in an unchanged state into the feces.[21]

Recent studies by Yoshimura et al.[22] on 2,4,3',4'-tetrachlorobiphenyl excretion in the rat have confirmed the above conclusion.   As can be seen from Table 4.7, the unchanged compound excreted into the feces during the first 4 days accounted for about 2.5% of the original dose, or on average about 0.6% of the dose per day. This result itself suggested that the gastrointestinal tract was an important site for the excretion of 2,4,3', 4'-tetrachlorobiphenyl in rats, since it had already been established that unchanged 2,4,3',4'-tetrachlorobiphenyl was not excreted into the bile.[20] Studies were made on the cumulative excretion rate of the unchanged compound in different parts of the gastrointestinal tract of rats, with the bile duct ligated after intravenous injection. It was found that about 1 hr after injection, unchanged compound (about 0.1% of the original dose) began to appear in the content of the small intestine but not in that of other parts. During periods of 3 and 6 hr after injection, the unchanged compound excreted into the small intestine accounted for about 0.2 and 0.3% of the dose, respectively, and trace amounts of 2,4,3',4'-tetrachloro-biphenyl were also detected in the contents of the cecum, colon and rectum during these periods. The results thus clearly suggested that the wall of the small intestine is a major site for the excretion of unchanged 2,4,3', 4'-tetrachlorobiphenyl in the rat.

From the toxicological point of view, this excretion route (i.e. through the small intestinal wall) is very important, since it may be common with that for the excretion of other compounds which are neutral, highly lipid-soluble and metabolically stable like PCB's. The route may also prove useful if it can be utilized clinically for promoting the excretion of PCB components from human tissues.

Concerning other modes of excretion of PCB's, those via the milk,[23] sebum,[2] and sputum[24] are known; however, their contribution to PCB elimination does not appear in general to be so great. Although considerable amounts of PCB's are known to be excreted into human milk, this case is of course limited to women suckling infants.

4.3 METABOLISM OF PCB's

It is well known that numerous foreign compounds, such as drugs, food additives, pesticides and industrial chemicals, once they have penetrated into the body of animals via various pathways, are metabolized to more polar derivatives.[25] The biotransformations involved tend to lead to an alteration of the original biological activity and to a rapid excretion of the foreign compounds.

PCB's, although they in general exhibit a high degree of biological and chemical stability, must also be transformed into more polar metabolites for effective excretion. Studies on PCB metabolism are therefore important in exploring the toxic properties of PCB's and in understanding the dynamic aspects of their pollution of the environment. Although such pollution of the environment by PCB's was recognized over a decade ago,[3] little was known until recently of the metabolic fate of PCB's in animals. This is in spite of the fact that many earlier investigators demonstrated a considerable difference in gaschromatographic patterns between commercial PCB mixtures and the residues found in animal tissues, with its obvious implication that individual components of commercial PCB preparations may be metabolized at different rates within the body of animals.

Grant et al.[12] suggested that in the rat, components of Aroclor 1254 with a lower chlorine content were metabolized to a greater extent than those with a higher chlorine content.   As mentioned above, they also reported that the tissue contents of the residues were significantly greater in rats with a carbon tetrachloride-damaged liver than in rats with a normal liver after the oral administration of Aroclor, indicating the liver as a major site of Aroclor 1254 metabolism.   Yoshimura et al.[9] concurrently investigated the distribution and elimination of PCB's in animals using ^{3}H-labeled Kanechlor 400, as described in section 4.2.   They obtained results similar to those of Grant et al., and also found additional evidence to suggest that the radioactivity eliminated in the feces of rats, although composed mainly of unchanged Kanechlor 400, also included phenolic metabolites in part, after the oral administration of Kanechlor 400.   This observation is consistent with the fact that the major metabolic pathway of aromatic compounds is their hydroxylation.[25]

It appears very difficult, however, to obtain more detailed information on either the quantitative or qualitative aspects of PCB metabolism, since commercial preparations are complex mixtures of various chlorobiphenyls.   For example, Kanechlor 400 may be separated into about 30 peaks using a gas chromatograph equipped with an electron capture detector (as shown in Fig. 4.1).   Sissons et al.[26] have also demonstrated that Aroclor 1254 can be resolved into more than 50 components by high resolution gas chromatography employing SCOT columns.

Peakall et al.[27] have pointed out that "we are still relatively unsophisticated in our PCB quantitation methodology and will continue to estimate only relative amounts of PCB's in field samples until we synthesize the individual PCB components commonly found in the ecosystem and are able to speak in terms of these individual peaks as we do for most pesticide residues".   Clearly, from the analytical, metabolic and toxicological viewpoints, there is a real need to isolate or synthesize the individual PCB components.

4.3.1   Isolation of individual components from commercial PCB preparations, and their chemical synthesis

Several investigators, [26–30] mainly concerned with analytical aspects, have attempted to isolate and characterize the individual isomers of commercial PCB's. In most of these studies, the major PCB constituents of Aroclors were characterized by high resolution nuclear magnetic resonance and mass spectrometry following separation by liquid-solid and/or gas-liquid chromatography. However, the structures of the separated components were not completely determined by comparison with the corresponding authentic compounds.

Saeki et al.[11] recently succeeded for the first time in isolating 6 of the main components of the PCB mixture, Kanechlor 400, in the pure crystalline state by an effective combination of preparative gas chromatography and fractional crystallization. (A typical gas chromatogram of Kanechlor 400, including the peaks of the isolated components, is shown in Fig. 4.1). These peaks were characterized as tetra- and pentachlorobiphenyls by ultraviolet absorption and mass spectrometry (Table 4.8). For deter-

TABLE 4.8

Physical Properties of the Main Components of Kanechlor 400 Isolated
(From Saeki et al.[11])

Compound	g_1	g_2	h	k	l	m
mp(°C)	120	95	138	174	95	103
MW	290	290	290	290	324	324
UV (nm)	253	250	250	259	253	250
(log ε)	(4.21)	(4.07)	(4.13)	(4.35)	(4.08)	(3.36)
Molecular formula	$C_{12}H_6Cl_4$	$C_{12}H_6Cl_4$	$C_{12}H_6Cl_4$	$C_{12}H_6Cl_4$	$C_{12}H_5Cl_5$	$C_{12}H_5Cl_5$

mination of their complete structures, chemical synthesis of the isolated PCB constituents was attempted by the coupling reactions of appropriate chlorine-substituted derivatives of aniline and benzene in the presence of amylnitrite, according to the method of Cadogan.[31] These reactions were performed only on a small scale and the products were examined by gas chromatography. Those reaction products showing identical retention times to each of the isolated components were then synthesized on a large scale by the more precise methods shown in Fig. 4.2. Thus, the complete structures for the g_1, g_2, h, k and m peaks were finally established as 2,4,3',4'-, 2,5,3',4'-, 2,3,4,4'-, 3,4,3',4'-tetrachloro- and 2,3,4,3',4'-pentachlorobiphenyl, respectively.

Hutzinger et al.[32] have studied the chemical synthesis of 23 individual PCB's by various methods, although some of them are unlikely to form components of commercial preparations. More recently, Webb et al.[29] have reported that 27 components of Aroclor 1221, 1242 and 1254 have been separated and identified by comparison of both their GLC retention times and infrared spectra with authentic samples. At present, more than 50 individual PCB's are supplied as commercial preparations, and utilized for biological and analytical studies.

Together with this work, tritium-labeled Kanechlor 400[9] was recently synthesized by the chlorination of biphenyl labeled with tritium at

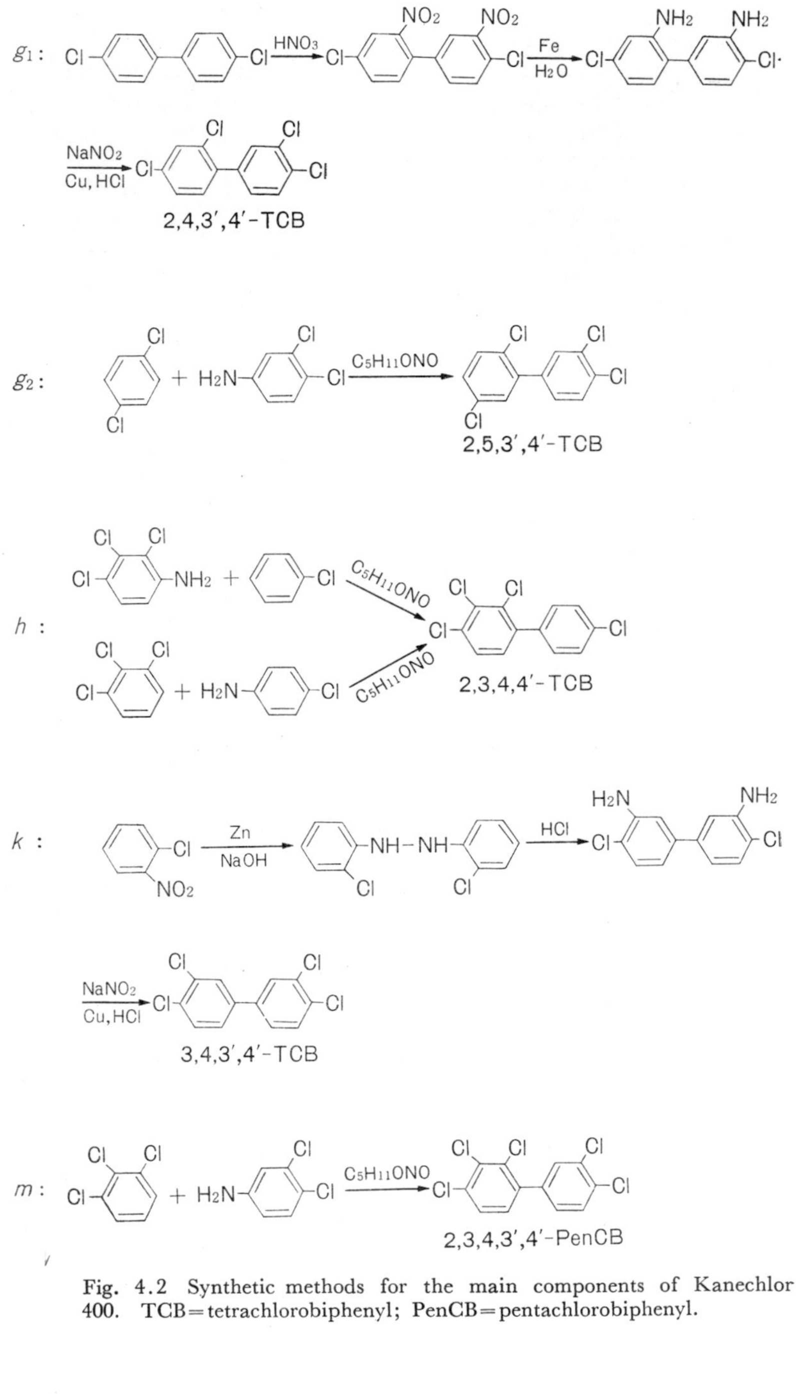

Fig. 4.2 Synthetic methods for the main components of Kanechlor 400. TCB=tetrachlorobiphenyl; PenCB=pentachlorobiphenyl.

specific positions, and Aroclor-^{36}Cl[33] was prepared by neutron irradiation of the commercial product, particularly to facilitate metabolic studies. A few papers are also available on the synthesis of individual PCB's labeled with tritium,[34] deuterium,[35] ^{36}Cl[35-37] and ^{14}C[38-40] at specific positions.

4.3.2 Metabolism of mono-, di- and trichlorobiphenyls

As mentioned above, it is recognized in general that one major metabolic reaction for aromatic compounds is hydroxylation. In this connection, biphenyl, the parent compound of PCB's has been shown to be metabolized mainly to 4-hydroxybiphenyl and to a lesser extent to 2-hydroxy- and 3,4-dihydroxybiphenyl in various animals.[41-43]

The first attempt to study the metabolism of a chemically defined chlorobiphenyl was that of Block et al.[42] with 4-chlorobiphenyl. In this case, the 4-chlorobiphenyl was shown to be metabolized to 4'-hydroxy-4-chlorobiphenyl and its glucuronide in rabbits.

Hutzinger et al.[21] recently examined the metabolic behavior of 4 Aroclor components with different chlorine contents (i.e. 4-, 4,4'-di-, 2,5, 2',5'-tetra- and 2,4,5,2',4',5'-hexachlorobiphenyl) in rats, pigeons and trout by means of gas chromatography and mass spectrometry. It was suggested that rats metabolized 4-chlorobiphenyl to monohydroxy and dihydroxy derivatives, whereas pigeons metabolized it only to monohydroxy-4-chlorobiphenyl. Although no evidence was obtained concerning the position of substitution, the monohydroxylated metabolite was also found in the case of 4,4'-dichlorobiphenyl in the rat urine and pigeon excreta. Trout, however, were unable to metabolize any of the PCB's investigated. Precise data regarding the rates of metabolism and excretion were not available in this study; however, it seemed more difficult to metabolize dichlorobiphenyl than the monochloro derivative, since large quantities of unchanged material were recovered in the case of dichlorobiphenyl, whereas in the case of monochlorobiphenyl only a little of the starting material could be recovered. The metabolic fates of tetra- and hexachlorobiphenyls are described in later sections.

Using several ^{14}C-labeled dichlorobiphenyls, Goto et al.[39,44,45] recently investigated their metabolism in rats after oral administration. As summarized in Fig. 4.3, all of these compounds were converted to the corresponding monohydroxy derivatives, being excreted mainly into the feces. It appeared that hydroxylation occurred predominantly at the p-position of the unsubstituted aromatic ring. However, in the case of dichlorobiphenyls; with chlorine on both aromatic rings, hydroxylation occurred at either the o- or p-position. In the case of 2,3-dichlorobiphenyl, about 15% of the total radioactivity fed to the rat was excreted as metabolites (mainly in the feces) during the first 5 days after oral administration.[44] In addi-

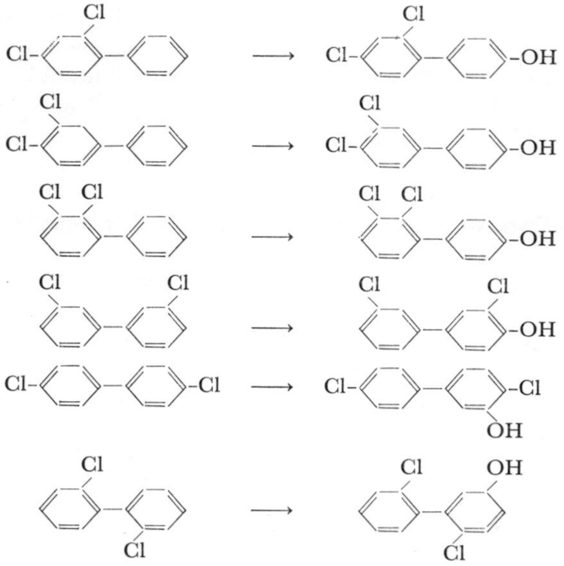

Fig. 4.3   Scheme for the metabolism of dichlorobiphenyls in rats.

tion, from data obtained by high resolution mass spectrometry, it was suggested that all the compounds were metabolized also to their dihydroxy derivatives.

The metabolism of 2,4,6-trichlorobiphenyl in the rat has been studied by the same investigators[39,45] using the ^{14}C-labeled compound. They found that it was converted to the 4'-hydroxy, 3',4'-dihydroxy and 3'-hydroxy-4'-methoxy and/or 4'-hydroxy-3'-methoxy derivatives after oral administration (Fig. 4.4). Hydroxy-methoxy derivatives have also been

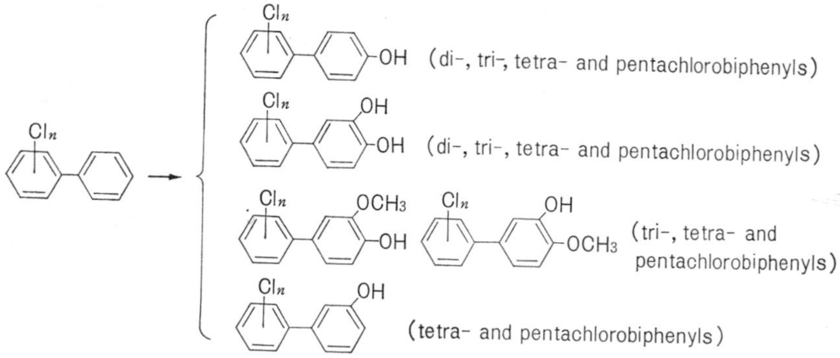

Fig. 4.4   Scheme for the metabolism of polychlorobiphenyls in rats.

identified as metabolites of biphenyl in the urine of rabbits in another study.[46]

4.3.3 Metabolism of tetrachlorobiphenyls

Tetrachlorobiphenyls are the principal components of the widely used commercial PCB's, such as Kanechlors (400 and 500) and Aroclors (1248 and 1254), and are also found in wildlife, etc. along with penta- and hexachlorobiphenyls.   In order to resolve the toxicological problems of PCB's, therefore, metabolic studies on these compounds must be promoted most.

As mentioned above, the metabolism of 2,5,2',5'-tetrachlorobiphenyl, one of the principal components of the Aroclors (1248 and 1254), has been investigated in animals by Hutzinger et al.[21]   Excreta from the animals were extracted and examined by gaschromatographic and mass spectrometric techniques.   The results indicated that this chlorobiphenyl was converted to a monohydroxylated metabolite in the rat and pigeon, but no hydroxy metabolites were detected in the case of the brook trout.   The same authors also reported that in the rat this metabolite could be detected only in the urine (not in the feces) after intraperitoneal injection.   In this study, however, the location of the hydroxyl group on the metabolite was not established.

Recently, Gardner et al.[47] have undertaken work to determine the complete structure of 2,5,2',5'-tetrachlorobiphenyl metabolites in rabbits. They found that animals fed with 2,5,2',5'-tetrachlorobiphenyl excreted three hydroxylated metabolites into their urine (Fig. 4.5), of which two were identified as 3-hydroxy- and 4-hydroxy-2,5,2',5'-tetrachlorobiphenyl by chromatographic and spectrometric comparison with synthetically pre-

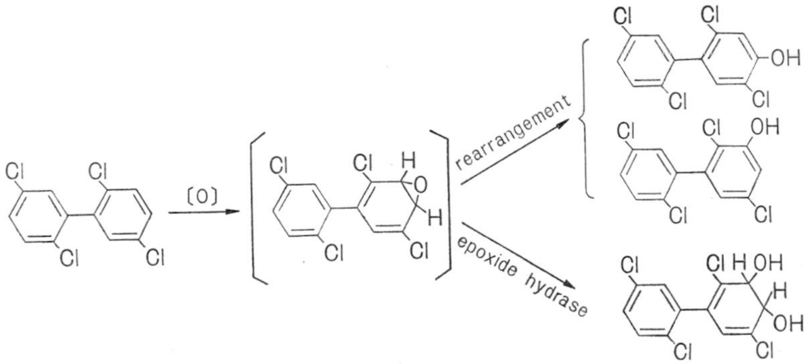

Fig. 4.5 Scheme for the metabolism of 2,5,2',5'-tetrachlorobiphenyl in rabbits.

pared compounds.   The third metabolite was shown to be *trans*-3,4-dihy-dro-3,4-dihydroxy-2,5,2′,5′-tetrachlorobiphenyl from its infrared and mass spectra and the infrared spectrum of its dehydration product. Based on these results, it was suggested that the arene oxide, 2,5,2′,5′-tetra-chlorobiphenyl-3,4-oxide, was their immediate precursor, and that this was transformed to dihydrodiol via hydration and to monohydroxy metabolites via rearrangement.   Monohydroxy metabolites could also be produced by dehydration of the dihydrodiol formed.

Apart from these studies, it has been indicated by Yoshimura and his co-workers,[17–20,48] while attempting to develop a treatment method for patients suffering from Kanechlor 400 intoxication, that certain tetra-chlorobiphenyls are metabolized to monohydroxy derivatives.   They reported that, together with large amounts of unchanged materials, several metabolites showing phenolic properties are excreted almost exclusively into the feces of rats which have been administered 2,4,3′,4′- and 3,4,3′,4′-tetrachlorobiphenyl orally, as shown in Fig. 4.6.   The major and minor

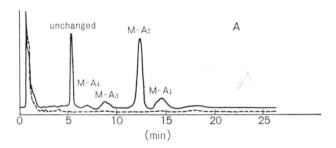

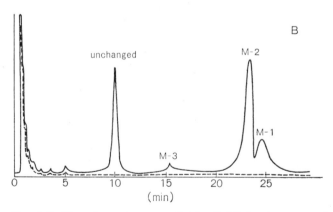

Fig. 4.6 Gas chromatograms of trimethylsilylated extracts from rat feces 3 days after oral administration of 2,4,3′,4′-(A) and 3,4,3′,4′-tetra-chlorobiphenyl (B) (solid lines); dotted lines show results for control experiments. (From Yoshimura *et al.*[17,18])

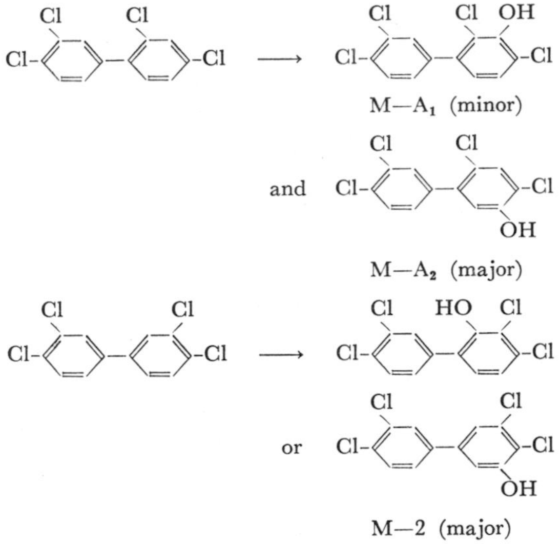

Fig. 4.7 Scheme for the metabolism of 2,4,3′,4′- and 3,4,3′,4′-tetra-chlorobiphenyls in rats.

metabolites of 2,4,3′,4′-tetrachlorobiphenyl, M–A$_2$ and M–A$_1$ respectively, were isolated from the feces and their structures characterized as mono-hydroxy derivatives by spectrometric techniques.   To elucidate the com-plete structure of these metabolites, 6 possible isomers of monohydroxy-2,4,3′,4′-tetrachlorobiphenyl were synthesized by a coupling reaction of appropriate diazotized dichloroanilines and dichlorophenols according to the method of Colbert et al.[49]   By comparing the spectral and chromato-graphic properties of these isomers with those of the isolated metabolites, M–A$_1$ (minor) and M–A$_2$ (major) were finally identified as 3-hydroxy- and 5-hydroxy-2,4,3′,4′-tetrachlorobiphenyl, respectively[48] (Fig. 4.7).

A major metabolite (M–2) isolated from the feces of rats which had been administered 3,4,3′,4′-tetrachlorobiphenyl orally was also character-ized as the monohydroxy derivative, and proved to be completely identical to one of the reaction products of the diazotized 3,4-dichloroaniline and 2,3-dichlorophenol.   This synthetic reaction should afford three isomeric products, 2- and 5-hydroxy-3,4,3′,4′-tetrachlorobiphenyl and 4-hydroxy-2,3,3′,4′-tetrachlorobiphenyl; thus, M–2 should be either 5- or 2-hydroxy-3,4,3′,4′-tetrachlorobiphenyl since the metabolic formation of 4-hydroxy-2,3,3′,4′-tetrachlorobiphenyl from 3,4,3′,4′-tetrachlorobiphenyl could be considered impossible[18] (Fig. 4.7).   The metabolites were excreted very slowly into the bile as free phenols, but not as conjugates.[20]   The excre-tion rate of 5-hydroxy-2,4,3′,4′-tetrachlorobiphenyl, the major metabolite of 2,4,3′,4′-tetrachlorobiphenyl, was about 10% of the dose over a period of 12 days from oral administration,[17] whereas the major metabolite of

3,4,3',4'-tetrachlorobiphenyl excreted during the first 14 days accounted for about 3% of the dose.[19]   The same metabolites were also excreted into the feces of mice injected with the above tetrachlorobiphenyls.[20]

Goto *et al.*[40,45] have reported in their series of studies that 2,3,5,6-tetrachlorobiphenyl may be metabolized to 4'- and 3'-hydroxy, 3',4'-dihydroxy, and 4'-hydroxy-3'-methoxy or 3'-hydroxy-4'-methoxy derivatives in rats (Fig. 4.4).   This metabolic patterns is very similar to that for unsubstituted biphenyl in rabbits.[46]   Most of the metabolites were excreted into the feces preferentially in the free form, although a small part was excreted into the urine exclusively as conjugates.

On the basis of the above reports by four different groups of workers, it is evident that tetrachlorobiphenyls are metabolized in animals essentially by the hydroxylation reaction.   However, it should be noted that the metabolic rate and excretion route of the metabolites may differ con-

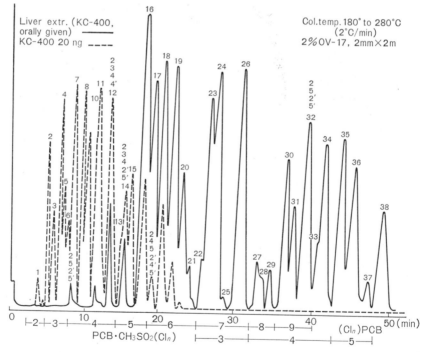

Fig. 4.8   Gas chromatograms of standard Kanechlor 400 (----) and liver extracts (——) of mice compelled to run after oral administration of Kanechlor 400.   (From Mio *et al.*[50])   The mice were fed on a diet containing 500 ppm Kanechlor 400 for 30 days, and prior to sacrifice were compelled to run in a rotating cage (6 rpm) for 1—3 days.   (The column temp. was raised continuously from 180 to 280°C at 2°C/min.)

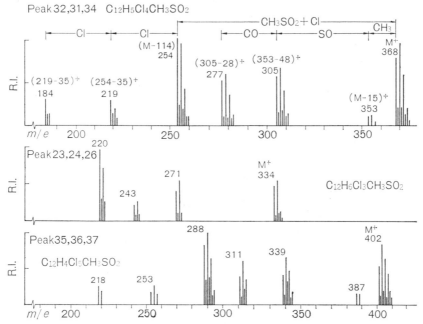

Fig. 4.9 Mass spectra of metabolites of accumulated Kanechlor 400 from the livers of mice compelled to run (experimental conditions as in Fig. 4.8). (From Mio *et al.*[50])

siderably according to the particular isomer, and particular animal species.

Quite recently, a very interesting finding concerning the metabolic fate of PCB's has been presented by Mio *et al.*[50]   In order to explore the metabolic behavior of accumulated PCB's in the body, mice were compelled to run for many hours in a rotating cage following the oral administration of 2,5,2′,5′-tetrachlorobiphenyl.   It was found that large amounts of this PCB were transferred from adipose tissue to the liver, from which a new type of metabolite was extracted with acetone.   This metabolite was analyzed by a combined gaschromatographic/mass spectrometric technique.   As shown in Figs. 4.8 and 4.9, the metabolite exhibited a longer retention time by gas chromatography and a molecular ion at $m/e = 368$ (original PCB+78) by mass spectrometry.   By high resolution mass spectrometry, it was proposed that the structure of this metabolite by either the methylsulfinate or methylsulfone derivative, both of which could be expressed as $C_{12}H_5Cl_4\text{-}SO_2CH_3$.   In similar experiments using PCB mixtures, i.e. Kanechlors (400, 500 and 600), it was confirmed that PCB's containing 3 to 5 chlorine atoms were converted to the corresponding compounds with a substituent SO_2CH_3 group, and that these were also ac-

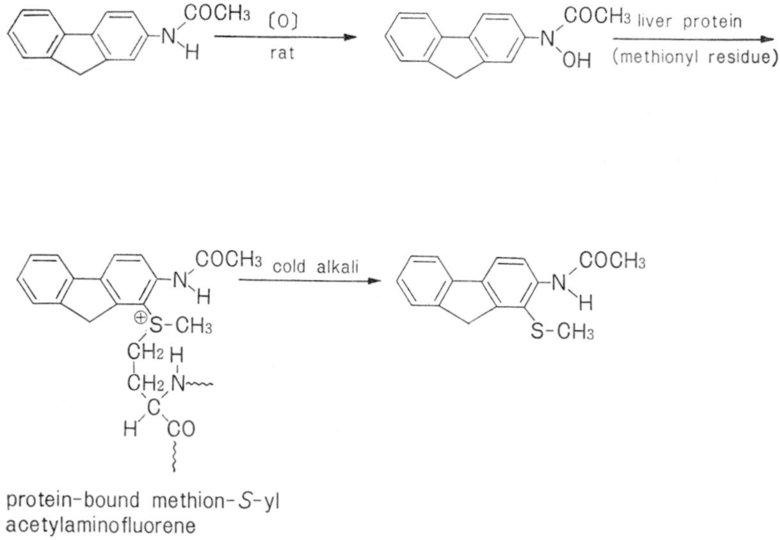

Fig. 4.10 Mechanism of 3-methylmercapto derivative formation from liver protein of rats given acetylaminofluorene.

cumulated in the livers of severely exercised mice (Figs. 4.8 and 4.9). The preferred structure from the metabolic viewpoint appeared to be the methylsulfone rather than the methylsulfinate; however, the complete structures of the PCB metabolites of this type (i.e. PCB–SO_2CH_3), including the positions of substitution by the SO_2CH_3 group, have not yet been determined.

In this connection, there are some interesting reports of liver proteins from rats which were given certain chemical carcinogens (such as dimethylaminoazobenzene and acetylaminofluorene) having yielded the corresponding 3-methylmercapto derivatives on treatment with cold alkali.[51] As indicated in Fig. 4.10, these methylmercapto derivatives are released from the methionyl residues of the proteins. Based on this evidence, it is thought that PCB metabolites containing the SO_2CH_3 group may be derived from the methylmercapto derivatives produced similarly to the above (see Fig. 4.12 below). It is well known that the sulfide group is easily transformed in biological reactions to sulfoxide and sulfone.[25] However, in order to confirm this assumption, additional evidence is still required.

4.3.4 Metabolism of penta- and hexachlorobiphenyls

Only a few studies have so far discussed the metabolic fate of chlorobiphenyls with more than 4 chlorine atoms per molecule.   One study was that carried out recently on pentachlorobiphenyl by Goto et al.[40,45]   However, their isomer was a rather special one, in which all the chlorine atoms were located on one side of the phenyl rings.   The metabolites identified are summarized above in Fig. 4.4.   In addition, Mio et al.[50] have presented evidence suggesting the formation of SO_2CH_3-derivatives of pentachlorobiphenyls in the liver of mice compelled to run, as well as of tetrachlorobiphenyls.

On the other hand, Yamamoto et al.[52] were unable to detect any hydroxylated metabolites in the excreta or tissues of rats in their metabolic studies of 2,3,4,3',4'-pentachlorobiphenyl, which is a component of Kanechlor 400.   This isomer exhibits a strong tendency to bind to liver constituents and has a high toxicity (see section 4).

Very recently, Jensen et al.[53] have reported in rats and mice that 2,4,5,2',4',5'-hexachlorobiphenyl, a chlorobiphenyl containing only isolated unsubstituted positions, is metabolized very slowly to 3-hydroxy-2,4,5,-2',4',5'-hexachlorobiphenyl (Fig. 4.11).   The total amount of metabolite

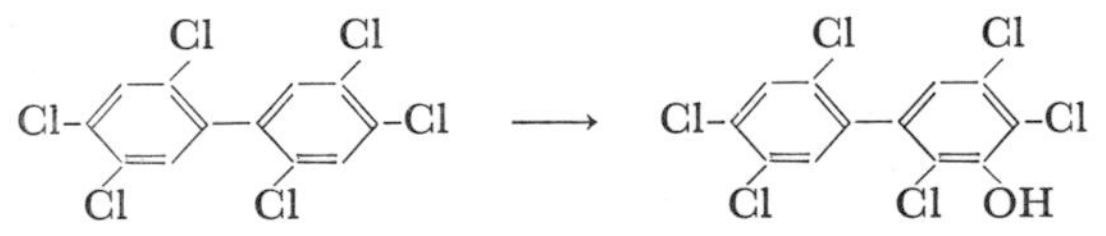

Fig. 4.11 Scheme for the metabolism of 2,4,5,2',4',5'-hexachlorobiphenyl in rats.

excreted into the feces over a 7-day period was calculated at about 1.3% of the oral dose, but no trace of this metabolite or its conjugates was found in the urine.   The metabolism of this chlorobiphenyl was first studied by Hutzinger et al.[21] using the rat, pigeon and trout, but no hydroxylated metabolites were detected in the excreta.

The observations described above thus suggest that most PCB components containing 4 or less chlorine atoms are essentially hydroxylated in mammals.   Certain isomers of penta- and hexachlorobiphenyls can, in addition, be metabolized, though very slowly.   In conclusion, the metabolism of chlorobiphenyls appears to become increasingly difficult as the number of chlorine atoms in the molecule increases.   It is also suggested that the vicinal hydrogen atoms of PCB's are preferred for metabolic breakdown of the molecule.[54]   Many investigations of the PCB residues found in man and animals tend to support this suggestion.   The hydroxylation step is thought to proceed via a reactive epoxide (arene oxide), which not only affords a phenol by rearrangement but also yields dihydrodiol by hydration with epoxide hydrase and glutathione conjugate with glutathione-S-epoxide transferase.   It is also probable that methylsulfone metabolites are formed through this epoxide intermediate, as shown in Fig. 4.12.   Such an ability for interacting with biopolymers such as pro-

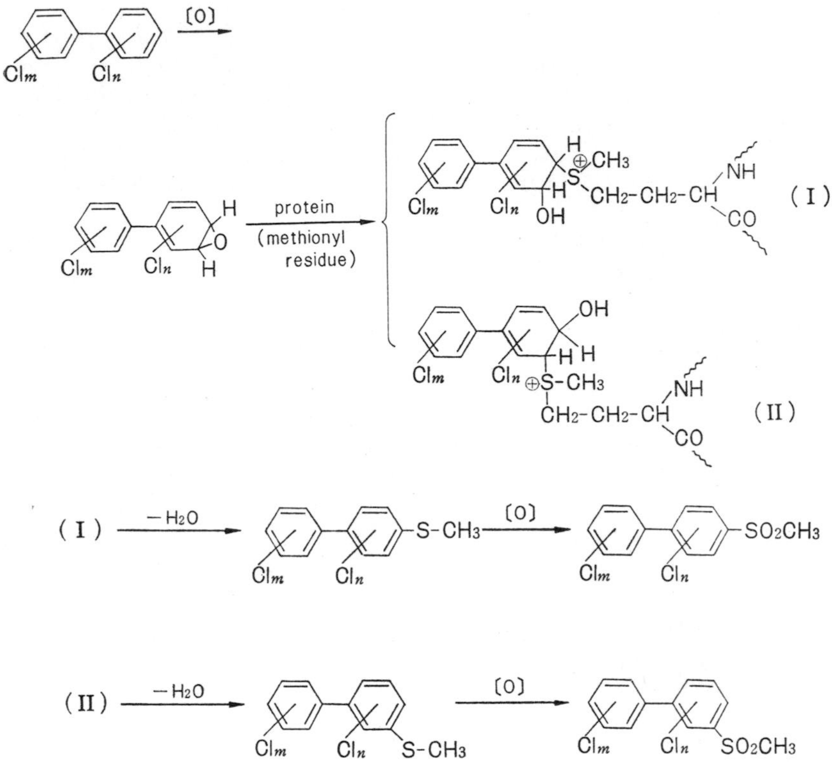

Fig. 4.12 Postulated mechanism for the formation of methylsulfone metabolites *in vivo*.

teins and nucleic acids, resembles that of dimethylaminoazobenzene or acetylaminofluorene, and is undoubtedly very important toxicologically.

4.4   TOXICOLOGICAL EVALUATION OF THE METABOLISM OF PCB'S

4.4.1   Toxicity of phenolic metabolites of PCB's

From the toxicological standpoint, the metabolism of foreign compounds is directed towards their inactivation or detoxication.   It is also true, however, that phenolic metabolites frequently show biological activities comparable to or even greater than the parent compounds.[55]   The important question therefore arises as to whether metabolites of PCB's

TABLE 4.9

Mortality and Lethal Dose of 2,4,3',4'-Tetrachlorobiphenyl and Its
Major Metabolite (M—A$_2$) in Mice (From Yamamoto et al.[48])

Dose (g/kg) (i. p.)[†1]	No. of dead mice				
	24 hr	48 hr	72 hr	96 hr	Total
2,4,3',4'-TCB					
0.35	0	0	0	0	0/8
0.70	0	0	0	0	0/8
1.05	0	0	0	0	0/8
1.40	0	0	0	0	0/8
1.75	0	0	1	1	2/8
2.10	0	0	1	3	4/8
2.45	0	0	2	3	5/8
3.15	1	1	3	2	7/8
	$LD_{50} = 2.15 (1.79—2.58)$ g/kg $(p < 0.05)$ [†2]				
M—A$_2$					
0.20	0	0	0	0	0/8
0.30	1	1	0	0	2/8
0.40	3	1	0	0	4/8
0.50	5	1	0	0	6/8
0.75	5	1	0	0	6/8
	$LD_{50} = 0.43 (0.34—0.55)$ g/kg $(p < 0.05)$ [†2]				

[†1] Each compound was injected intraperitoneally into male CF ♯1 strain mice after being dissolved in soybean oil (0.2—0.5 ml). The animals (8 in each group) were housed in a room maintained at 20°C, and the numbers of dead mice were counted 24, 48, 72 and 96 hr after administration.

[†2] The LD_{50}'s were determined by the method of Litchfield–Wilcoxon.[56]

exhibit greater or lesser toxicity than the parent PCB's.

In order to answer this question, Yamamoto et al.[48] initiated the series of metabolic studies on individual PCB components described above, and identified the major metabolite of 2,4,3',4'-tetrachlorobiphenyl as the 5-hydroxyderivative. In subsequent work, they determined the acute lethal dosage by intraperitoneal injection into male mice. The results showed that the LD_{50}'s of 2,4,3',4'-tetrachlorobiphenyl and its metabolite were 2.15 and 0.43 g/kg, respectively, suggesting that the acute toxicity of 2,4,3',4'-tetrachlorobiphenyl might be attributable to the production of the phenolic metabolite within the body (Table 4.9). If this were true, then the lethal effect of 2,4,3',4'-tetrachlorobiphenyl would presumably increase in animals pretreated with phenobarbital or 3-methylcholanthrene, since such pretreatment is thought to accelerate the hydroxylation of foreign compounds.[57] In fact, the acute toxicity of this chlorobiphenyl increased slightly in mice pretreated with 3-methylcholanthrene. It was also found that following such pretreatment, rats excreted 10 times as much of the hydroxy metabolite into the bile as did untreated rats. On the other hand, phenobarbital pretreatment unexpectedly led to a decrease in the lethal effect (LD_{50} over 3 g/kg), in spite of a two-fold increase in the fecal excretion of 5-hydroxy-2,4,3',4'-tetrachlorobiphenyl.[58] However, further studies suggested that phenobarbital pretreatment might actually accelerate the formation of the toxic metabolite slightly, but also promote its stronger biliary excretion.   The precise mechanism of such increased excretion is unknown at present, although a simple acceleration of bile flow caused by phenobarbital (but not by 3-methylcholanthrene) pretreatment is one possibility.

On the basis of these findings, it appears probable that PCB toxicity may be attributable at least in part to phenolic metabolites produced within the body.   Recently, Brodie et al.[59] have reported that single doses of halogenobenzenes (1 ml/kg) injected intraperitoneally into rats can induce massive necrosis of the centrolobular regions of the liver.   They suggested that an epoxide produced in aromatic hydroxylation as a labile intermediate may be responsible for such hepatic necrosis.   Clearly, this hypothesis could be extended to PCB toxicity, although no evidence has yet been obtained for epoxide formation in the metabolism of PCB's, except for 2,5,2',5'-tetrachlorobiphenyl (see section 4. 3).

In relation to metabolically-induced PCB toxicity also, attention should be given to the possibility that chlorinated dibenzofurans, i.e. the very toxic compounds pointed out by Vos et al.,[60] may be formed.   That is to say, certain metabolites hydroxylated at the 2-(6-) or 2'-(6'-) position of the PCB molecule, particularly in the case of PCB's possessing a chlorine atom at the 2'-(6'-) or 2-(6-) position, should give dibenzofuran derivatives by intramolecular dehydrochlorination, as shown in Fig. 4.13.

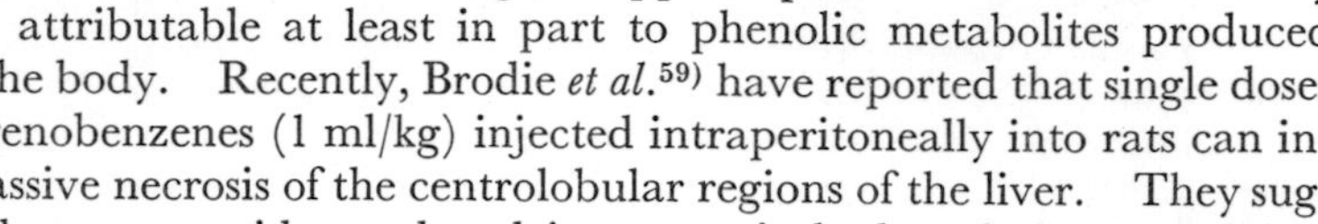

Fig. 4.13 Possible mechanism for the formation of polychlorinated dibenzofurans in the animal body.

4.4.2 Toxicity of 2,3,4,3',4'-pentachlorobiphenyl

During the course of studies on the metabolism of the Kanechlor component, 2,3,4,3',4'-pentachlorobiphenyl, Yamamoto et al.[52] found this compound to show a very severe toxicity.   It caused a marked decrease in body weight, disappearance of fat from adipose tissues, and liver necrosis, in rats after a single oral dose of 2 mg per animal (about 160 g).   Two of the 3 rats used in the experiment died after 2 weeks.   The intraperitoneal LD_{50}'s in mice were determined as 0.65 and 0.40 g/kg based on the number of deaths within 7 and 14 days, respectively, suggesting that mice may exhibit a lower susceptibility to this chlorobiphenyl than rats.

In no case were hydroxylated metabolites detected in the excreta or carcasses of animals given 2,3,4,3',4'-pentachlorobiphenyl.   It is very interesting to note, however, that when this chlorobiphenyl was incubated with rat liver homogenate in the presence of NADPH and oxygen, a considerably low recovery of unchanged compound was observed, and the recovery rate fell with increase in incubation time.[52] Such low substrate recovery was not observed in the case of incubation with preheated homogenate or with 2,4,3',4'-tetrachlorobiphenyl, which did not exert such a severe toxicity as 2,3,4,3',4'-pentachlorobiphenyl.

Based on these results, it may be speculated that some reactive intermediates formed metabolically may bind tightly to liver components and so cause severe hepatotoxicity. One of such possible intermediates is the epoxide suggested by Brodie et al.[59] to be an intermediate causing liver necrosis on the administration of halogenobenzenes. Another is a specific radical metabolite, also formed with liver microsomes, which has been suggested by several workers[61,62] to be the trigger of carbon tetrachloride-induced hepatotoxicity.

Although further studies are necessary to decide which mechanism is the more important, the latter at present affords a better explanation since no hydroxy metabolite has yet been detected in animals after 2,3,4,-3',4'-pentachlorobiphenyl administration. It should be noted also that possible contamination of the pentachlorobiphenyl sample with chlorinated dibenzofurans or benzodioxins has been ruled out by very careful gas-chromatographic examinations.

REFERENCES

1. F. J. Biros, A. C. Walker and A. Medbery, Bull. Environ. Contam. Toxicol., 5, 317 (1970).
2. Reports on the Study of "Yusho", No. 1, Fukuoka Acta Med., 60, 403 (1969).
3. S. Jensen, New Scientist, 32, 612 (1966).
4. R. W. Risebrough, P. Reiche, D. B. Peakall, S. G. Herman and M. N. Kirven, Nature, 220, 1098 (1968).
5. E. L. McCune, J. E. Savage and B. L. O'Dell, Poultry Sci., 41, 295 (1962).
6. L. Schwartz, Am. J. Pub. Health, 26, 586 (1936).
7. C. K. Drinker, M. F. Warren and G. A. Bennett, J. Ind. Hyg. Toxicol., 19, 283 (1937).
8. S. Ikegami, K. Kawamoto, Y. Kashida, K. Enogaki and S. Baba, Radioisotopes, 20, 65 (1971).
9. H. Yoshimura, H. Yamamoto, J. Nagai, Y. Yae, H. Uzawa, Y. Ito, A. Notomi, S. Minakami, A. Ito, K. Kato and H. Tsuji, Fukuoka Acta Med., 62, 12 (1971).
10. H. Yoshimura and M. Oshima, ibid., 62, 5 (1971).
11. S. Saeki, A. Tsutsui, K. Oguri, H. Yoshimura and M. Hamana, ibid., 62, 20 (1971).
12. D. L. Grant, W. E. J. Phillips and D. C. Villeneuve, Bull. Environ. Contam. Toxicol., 6, 102 (1971).
13. R. T. Williams, P. Millburn and R. L. Smith, Ann. N. Y. Acad. Sci., 123, 110 (1965).

14. P. A. Shore, B. B. Brodie and C. A. M. Hogben, *J. Pharm. Exptl. Therap.*, **119**, 361 (1957).
15. E. L. Noach, D. M. Woodbury and L. S. Goodman, *ibid.*, **122**, 301 (1958).
16. C. E. Harrison Jr., R. O. Branoenburg, P. Ongley, A. L. Orivis and C. A. Owen Jr., *J. Lab. Clin. Med.*, **67**, 764 (1966).
17. H. Yoshimura, H. Yamamoto and S. Saeki, *Chem. Pharm. Bull. (Tokyo)*, **21**, 2231 (1973).
18. H. Yoshimura and H. Yamamoto, *ibid.*, **21**, 1168 (1973).
19. H. Yoshimura and H. Yamamoto, *Fukuoka Acta Med.*, **65**, 5 (1974).
20. H. Yoshimura, H. Yamamoto and H. Kinoshita, *ibid.*, **65**, 12 (1974).
21. O. Hutzinger, D. M. Nash, S. Safe, A. S. W. DeFreitas, R. J. Norstrom, D. J. Wildish and V. Zitko, *Science*, **178**, 312 (1972).
22. H. Yoshimura and H. Yamamoto, *Bull. Environ. Contam. Toxicol.*, in press.
23. N. Isono and K. Fujihama, *Kagaku* (Japanese), **42**, 396 (1972).
24. T. Kojima, *Fukuoka Acta Med.*, **62**, 25 (1971).
25. D. V. Parke, *The Biochemistry of Foreign Compounds*, Pergamon, 1968.
26. D. Sissons and D. Welti, *J. Chromatogr.*, **60**, 15 (1971).
27. D. B. Peakall and J. L. Lincer, *BioSci.*, **20**, 958 (1970).
28. J. W. Rote and P. G. Murphy, *Bull. Environ. Contam. Toxicol.*, **6**, 377 (1971).
29. R. G. Webb and A. C. McCall, *J. Assoc. Offic. Anal. Chem.*, **55**, 746 (1972).
30. S. N. Hirwe, R. E. Borchard, L. G. Hansen and R. L. Metcalf, *Bull. Environ. Contam. Toxicol.*, **12**, 138 (1974).
31. J. I. G. Cadogan, *J. Chem. Soc.*, **1962**, 4257.
32. O. Hutzinger, S. Safe and V. Zitko, *Bull. Environ. Contam. Toxicol.*, **6**, 209 (1971).
33. D. L. Stalling and J. N. Huckins, *J. Assoc. Offic. Anal. Chem.*, **54**, 801 (1971).
34. O. Hutzinger and S. Safe, *Bull. Environ. Contam. Toxicol.*, **7**, 374 (1972).
35. O. Hutzinger, C. Pothier and S. Safe, *J. Assoc. Offic. Anal. Chem.*, **55**, 753 (1972).
36. T. Migita, R. Ito, K. Tori and O. Shimamura, *Yuki Gosei Kagaku Kyokai-shi* (Japanese), **19**, 609 (1961).
37. T. Migita, N. Morikawa and O. Shimamura, *Bull. Chem. Soc. Jap.*, **36**, 980 (1963).
38. G. Sundström, *Bull. Environ. Contam. Toxicol.*, **11**, 39 (1974).
39. M. Goto, K. Sugiura, M. Hattori, T. Miyagawa and M. Okamura, *Chemosphere*, (5), 227 (1974).
40. M. Goto, K. Sugiura, M. Hattori, T. Miyagawa and M. Okamura, *ibid.*, (5), 233 (1974).
41. H. D. West, J. R. Lawson, I. H. Miller and G. R. Mathura, *Arch. Biochem. Biophys.*, **60**, 14 (1956).
42. W. D. Block and H. H. Cornish, *J. Biol. Chem.*, **234**, 3301 (1959).
43. P. Raig and R. Ammon, *Arzneim. -Forsch.*, **20**, 1266 (1970).
44. M. Goto, K. Sugiura, M. Hattori, T. Miyagawa and M. Okamura, *New Methods in Environ. Chem. and Toxicol.*, *Abstracts, Int. Symp. Ecolog. Chem.*, *Susono, Japan*, 1973, p. 299.
45. M. Goto, K. Sugiura and M. Hattori, *Abstracts, Sixth Symp. Drug Metabolism and Action, Tokyo*, 1974, p. 34.
46. P. Raig and R. Ammon, *Arzneim. -Forsch.*, **22**, 1399 (1972).
47. A. M. Gardner, J. T. Chen, J. A. G. Roach and E. P. Ragelis, *Biochem. Biophys. Res. Commun.*, **55**, 1377 (1973).
48. H. Yamamoto and H. Yoshimura, *Chem. Pharm. Bull. (Tokyo)*, **21**, 2237 (1973).
49. J. C. Colbert and R. M. Lacy, *J. Am. Chem. Soc.*, **68**, 270 (1946).
50. T. Mio, K. Sumino and T. Mizutani, *Abstracts, First Symp. Environ. Toxicol.*, *Osaka*, 1974, p. 23.
51. J. A. Miller, *Cancer Res.*, **30**, 559 (1970).
52. H. Yamamoto, H. Yoshimura, M. Fujita and T. Yamamoto, *Abstracts, First Symp. Environ. Toxicol.*, *Osaka*, 1974, p. 24.

53. S. Jensen and G. Sundström, *Nature*, **251**, 219 (1974).
54. E. Schulte and L. Acker, *Naturwissenschaften*, **61**, 79 (1974).
55. I. M. Fraser and E. S. Vesell, *J. Pharm. Exptl. Therap.*, **162**, 155 (1968).
56. J. T. Litchfield Jr. and F. Wilcoxon, *ibid.*, **96**, 99 (1949).
57. A. H. Conney, *Pharm. Rev.*, **19**, 317 (1967).
58. H. Yoshimura, *Farumashia* (Japanese), **7**, 786 (1972).
59. B. B. Brodie, A. K. Cho, G. Krishna and W. D. Reid, *Ann. N. Y. Acad. Sci.*, **179**, 11 (1971).
60. J. G. Vos, J. H. Koeman, H. L. Van der Maas, M. C. ten Noever de Brauw and R. H. de Vos, *Food Cosmet. Toxicol.*, **8**, 625 (1970).
61. T. F. Slater, *Nature*, **209**, 36 (1966).
62. E. S. Reynolds, *J. Pharm. Exptl. Therap.*, **155**, 117 (1967).

5

The Pathology of Yusho

Masahiro KIKUCHI
and Yoshito MASUDA

5.1 INTRODUCTION

As described in earlier chapters, the occurrence of patients with acne-form eruptions and skin pigmentation was noticed in northern Kyushu, Japan in the early summer of 1968. It was soon determined by the Study Group for Yusho that the lesions were induced by the accidental intake of Kanechlor 400 (chlorinated biphenyls), which had contaminated rice oil during its refining process in about February, 1968.[1,2] Recently, the deleterious effects of environmental pollution due to industrial products have been emphasized, and strong warnings against PCB pollution have been issued. In the present outbreak, the Kanechlor 400 was composed mainly of tetrachlorobiphenyls, so the lesions caused by its intake were carefully watched to know the possible influence of PCB's in the world.

There are few reports on pathological studies of PCB poisoning in man, except for those on direct cutaneous changes caused by contact with industrial materials containing PCB. The present chapter describes our histopathological findings and PCB gas-chromatographic analysis of visceral organs in autopsy cases, as well as our biopsy findings on the skin of Yusho patients.

5.2 AUTOPSY FINDINGS OF YUSHO PATIENTS

Up to May, 1974, 1122 Yusho patients had been discovered and 24 had died. Five of the dead patients and one other stillborn baby were autopsied. The essential clinical and autopsy findings were as follows:

CASE 1: 13-yr-old male; a schoolboy

Since spring 1967, his family (5 persons) had been using about 1.8 l of rice oil per month. In about July 1968, he noticed acne-like cutaneous eruptions and pigmentation on his face, head, chest and abdomen. Examination showed that abnormal pigmentation of the conjuctiva and hypersecretion of the meibomian glands were also severe. The boy was treated with several vitamin preparations, and the cutaneous changes considerably improved, leaving scars. On July 8, 1968, he died suddenly, coincident with the abrupt onset of a sensation of squeezing behind and to the left of the sternum, after having a run. No abnormal data were observed from blood, chemical analysis and other laboratory tests.

On autopsy he was found to be a well-nourished boy, but with numerous acneform eruptions on the face, axillae and abdominal wall. Pigmentation and flattening of the finger- and toe-nails were seen, and several soybean-sized palpable tumors were found on the left neck and axilla. The blood was not coagulated and diffuse petechial hemorrhage was seen on the pericardium. Histologically, the skin showed follicular hyperkeratosis and hyperpigmentation. The esophageal glands had dilated ducts with multilayered epithelium. Atrophy of the adrenal cortex was conspicuous.

Autopsy diagnosis: acneform eruptions and hyperpigmentation of the skin. Hypoplastic aorta. Pulmonary hemorrhage and edema. Petechial hemorrhage of the pericardium. Acute tracheobronchitis. Hyperplasia of lymphoid tissue in the lymph nodes, intestine and tonsils. Atrophic adrenal cortex.

CASE 2: 25-yr-old male; a truck driver[3]

His family (6 persons) had been using about 1.8 l of rice oil per month for several years. In summer 1968, he and his family noticed severe acnelike eruptions on the face, neck and chest, and he was treated for Yusho from October, 1968. In June, 1969, he was admitted to a hospital suffering from severe abdominal pain. Under suspicion of a perforated duodenal ulcer, an operation was performed. No perforation was found, but there was marked fibrous adhesion in the upper abdominal cavity, especially around the duodenum. Three weeks after the operation, on June 9, 1969, he died of acute heart failure.

On autopsy, marked fibrinofibrous pericarditis and multiple small areas of myocardial fibrosis or necrosis with myofibrillar degeneration and lymphocytic infiltrates were scattered in both the ventricular and atrial walls.   The dermal changes were almost identical with those of chloracne. Besides epidermal and follicular changes, hyperkeratosis and cystic dilatation of the sweat glands of the axillae were found.   Cystic dilatation and hyperplasia of the excretory duct epithelium of the esophageal gland were also detected.   The liver was congested but no necrotic changes were noticed.   Typical findings are illustrated in Figs. 5.1–5.6.

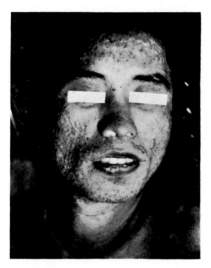

Fig. 5.1   Autopsy Case 2.   Acneform eruptions and pigmentation of the face.

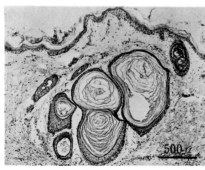
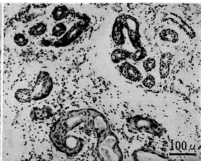

Fig. 5.2   Autopsy Case 2. Axilla. Hyperkeratosis and dilatation of the hair follicles (H–E stain, ×30).

Fig. 5.3   Autopsy Case 2. Axilla. Hyperplasis of the duct epithelium of the sweat glands (H–E stain, ×108).

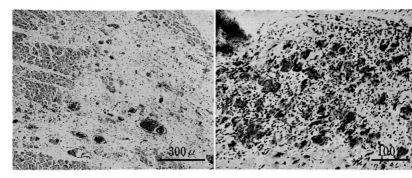

Fig. 5.4 Autopsy Case 2. Heart.
Small fibrou focus (H–E stain, ×62).

Fig. 5.5 Autopsy Case 2. Heart. Basophilic myofibrillar degeneration and inflammatory infiltrates (H–E stain, ×160).

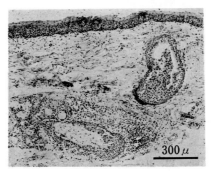

Fig. 5.6 Autopsy Case 2. Esophagus. Hyperplasis and dilatation of
the duct of the esophageal gland (H–E stain, ×61).

Autopsy diagnosis: multiple acneform eruptions with scars on the face,
neck, chest, axillae and scrotum.   Hyperpigmentation of the skin of the
face, hand and nail-bed.   Fibrinofibrous pericarditis (pericardial fluid,
50 ml). Multiple small fibrous or necrotic foci in the myocardium. Hyperplasia of the duct epithelium of the esophageal gland.   Fibrous adhesion
between the diaphragm and liver or spleen.   Ascites: 150 ml.   Pleural
fluid: left, 50 ml; right, 50 ml.   Pulmonary edema and congestion.   Dilatation of the right heart.   Postoperative state of gastrojejunostomia.

CASE 3:   73-yr-old male[3]
      At age 66 he received treatment for polyarthritis, and at age 69 treat-
ment for empyema.   In July, 1968, he was admitted to a hospital suffering
from upper abdominal pains, general malaise and anemia.   Acneform e-
ruptions were seen on his face, axillae and scrotum at that time.   A gastric ulcer and brochial asthma were found and he was treated with ap-

propriate drugs. At the end of October, 1969, he suddenly noticed a severe abdominal pain, at which time the liver edge was palpable 4 finger breadths below the right costal margin. The symptoms of cardiac failure increased and he died on November 7, 1969.

On autopsy, severe generalized amyloidosis was found. Skin changes such as acneform eruptions were seen on the face, axillae and acrotum. The duct of the esophageal gland showed cystic dilatation and hyperplasia of the epithelium.

Autopsy diagnosis: generalized amyloidosis (liver, spleen, tongue, heart, gastrointestinal tract, lungs and thyroid glands). Hyperpigmentation and acneform eruptions on the face. Chronic bronchitis and bronchial ectasia. Pulmonary edema. Ascites: 50 ml. Pleural fluid: left, 200 ml; right, 400 ml. Fibrinous pericarditis (300 ml). Pleural adhesion, bilateral. Severe generalized anemia. Soybean-sized myxoma at the left auricle of the heart.

CASE 4: 48-yr-old female; a housewife[4]

The patient had been suffering from skin eruptions (especially on the face) since May, 1968. She noticed the symptoms of liver disturbance in January, 1969, and was admitted to a hospital on March 13, 1969. Based on clinical findings and laboratory examinations, a diagnosis of liver cirrhosis was established. After admission, she was treated with liver-protecting drugs, but her condition did not improve. On December 29, 1970, she died after the onset of severe abdominal pain.

On autopsy, her enlarged liver showed multilobular cirrhotic changes and many nodules of hepatic cell carcinoma, some of which had associated necrosis and were covered with blood coagula. The cause of death was considered to be liver cell carcinoma rupture, with hemorrhage in the abdominal cavity. Mild skin changes were found on her face, neck and axillae. Histological examination revealed some dilated hair follicles with keratotic plugs.

Autopsy diagnosis: liver cirrhosis with hepatocellular carcinoma and pulmonary metastases. A few acneform lesions on the face, neck and axillae. 300 ml of hemoascites with 310 g of intraperitoneal blood coagula. Splenomegaly (200 g). Pleural fluid: right, 500 ml.

CASE 5: 46-yr-old male; a white-collar worker[5]

In summer 1968, he was diagnosed as a Yusho patient based on observed dermal changes. After April, 1972, he suffered from a persistent cough and general fatigue. On May 3, 1972, he began fast treatment to improve his skin symptoms and other general manifestations, but his condition worsened. He noticed high fever on May 12 and was admitted to the University Hospital as an emergency case on May 16. At the time of admission, severe cardiac insufficiency was already present and he died 2 hr after the admission.

On autopsy, acneform eruptions were found on his face, chest, axillae, and scrotum.   Histologically, typical hyperkeratosis and dilatation of the hair follicles were present.   The heart was 400 g in weight and multiple small necrotic foci with severe calcification were diseminated in the myocardium (Fig. 5.7).   Similar metastatic calcification was found in the kidneys, lungs, liver and mucosa of the stomach.   The examined bone showed rarefaction of the bone trabeculae, with prominent peritrabecular fibrosis (Fig. 5.8).   No hyperplastic or adenomatous changes were seen in the parathyroid glands.

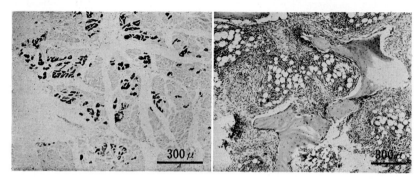

Fig. 5.7 Autopsy Case 5. Heart. Calcium deposition in myocardial fibers (Von Kossa's silver technique, ×61).

Fig. 5.8 Autopsy Case 5. Sternum. Resorption of bone trabeculae and peritrabecular fibrosis (H–E stain, ×48).

Autopsy diagnosis: osteitis fibrosa generalisata. Metastatic calcification of the heart, liver, lungs and gastric mucosa.   Myocardial hypertrophy with multiple necrotic foci.   Acneform eruptions on the face, chest, axillae and scrotum.

CASE 6:   stillborn[6)]
The mother of the stillborn, a 25-yr-old Japanese, had used rice oil continuously during her gestation period from January to October, 1968, and was diagnosed as a Yusho patient towards the end phase of her pregnancy based on acneform eruptions on her face and thigh (Fig. 5.9).   On October 24, 1968, she delivered a stillborn weighing 2600 g.

On autopsy, the skin of the stillborn showed a diffuse darkish brown color due to the presence of abundant melanin pigment similar to that of a negro (Figs. 5.10–5.12).   No acneform eruptions were noticed, but histologically, hyperkeratosis, atrophy of the epidermis and cystic dilatation of hair follicles with keratotic plugs were observed, especially in the head (Fig. 5.12).   Marked hyperemia of all organs, atelectasis of the lungs and slight hemorrhagic diathesis were also found.

Autopsy diagnosis: Stillborn. Hyperpigmentation of the skin, with keratotic follicular plugs. Generalized congestion. Slight hemorrhagic diathesis.

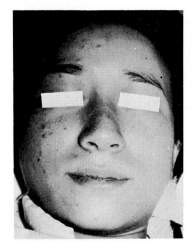

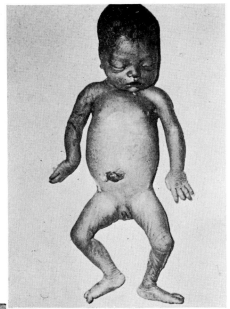

Fig. 5.9 Mother of Autopsy Case 6. Acneform eruptions on the cheek and forehead.

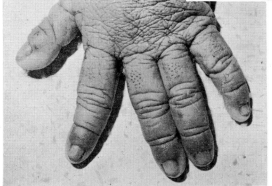

Fig. 5.10 Autopsy Case 6. Stillborn. Dark brownish pigmentation of the skin.

Fig. 5.11 Autopsy Case 6. Stillborn. Dark brownish pigmentation of the hand and nail-beds.

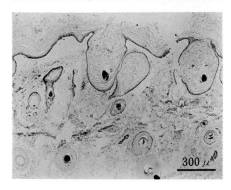

300 μ

Fig. 5.12 Autopsy Case 6. Head. Increased melanin pigment and hyperkeratosis of the epidermis, and keratotic plugs in the hair follicles (Masson-Fontana technique for melanin, ×48).

TABLE 5.1
Clinical Data for Autopsy Patients

Case	Age (yr)	Sex	Duration (months)	Severity†	Dose of rice oil	Date of death	Chief clinical diagnosis	Miscellaneous
1	13	M	12	Ⅲ	1.8 l/m. (5 persons)	July 8, 1969	Sudden death	No abnormal laboratory data
2	25	M	12	Ⅳ	1.8 l/m. (6 persons)	July 9, 1969	Acute abdominal pain Duodenal ulcer	Leukocytosis before death
3	73	M	16	Ⅱ		Nov. 7, 1969	Chronic heart failure	Hepatomegaly Increased BSR
4	48	F	29	Ⅰ		Dec. 29, 1970	Liver cirrhosis with carcinoma	Disturbance of liver function
5	46	M	46	Ⅲ	Kanechlor 1.6 g	May 16, 1972	Acute heart failure	Fast treatment Increased serum Ca
6	0	M	4 (mother)	Ⅲ (mother)	(mother)	Oct. 24, 1968	Stillborn	Coiling of the umbilical cord

† The severity of the lesions was divided into 4 groups, Grade IV being the severest and Grade I being the mildest.

TABLE 5.2
Autopsy Findings for Yusho Patients

Case	Age (yr)	Sex	Chief autopsy diagnosis	Skin lesion distribution	Pigmentation	Follicular hyperkeratosis	Hyperplasis of esophageal duct epithelium	Cause of Death
1	13	M	Liquor sanguinus	considerably wide	mild	moderate	absent	Acute heart failure
2	25	M	Multiple small fibrous or necrotic foci in the myocardium	very wide	moderate	severe	moderate	Acute heart failure
3	73	M	Generalized amyloidosis	face	very mild	moderate	moderate	Heart failure
4	48	F	Liver carcinoma and cirrhosis	face	very mild	mild	absent	Rupture of liver carcinoma
5	46	M	Osteitis fibrosa generalisata with metastatic calcification	face	very mild	moderate	mild	Heart failure
6	0	M	Stillborn Brown baby	very wide	severe	moderate	absent	Coiling of the umbilical cord

Summary of autopsy cases :   the principal clinical and autopsy findings are summarized in Table 5.1 and 5.2.   The autopsies were performed
between 12 and 46 months after the first appearance of skin lesions, except
for the case of the stillborn, which was autopsied 4 months after the first
appearance of skin lessions in the mother.   The total amounts of Kanechlor ingested were not known exactly, although in Case 5 it was suspected
that about 1.6 g had been ingested.   Histopathological changes which
were probably due to Kanechlor included hyperkeratosis of the hair follicles and increased melanin pigment in the basal layer of the epidermis:
apart from this, there were no definitive changes in the visceral organs,
although in 3 of 5 cases (except the stillborn) proliferation of the duct
epithelium of the esophageal glands was seen.

Experimental and electron microscopic studies have revealed changes
in the liver,[7,8] pericardium,[9] and heart.[10]   However, the autopsy cases
here showed no definitive changes in the liver, except for one instance of
liver cirrhosis with carcinoma (Case 4) and one instance of amyloid deposition (Case 3).   These changes, however, were probably unrelated to the
chlorobiphenyl poisoning since their first occurrence was suspected to have
been before the actual intake of Kanechlor.   As for changes in the heart,
there was one instance of multiple small fibrous or necrotic foci, with basophilic changes of the muscle fibers (Case 2) and another of metastatic
calcification (Case 5).   The basic nature of both these conditions was not
determined.

5.3   Cutaneous Changes in Yusho Patients[11]

The main clinical findings of Yusho patients concern dermatosis.   As
described in detail in Chapter 7, acneform eruptions, prominent enlargement and elevation of the follicular openings, pigmentation and flattening
of the nails, and pigmentation of the skin are considered to major cutaneous changes.[12]   Skin biopsy was thus performed on Yusho patients at 24
sites in 18 patients between 20 days and 5 months after the appearance of
cutaneous lesions.   The patients examined were aged 3 to 54, and the examined sites were 10 in the cheek, 5 in the neck, 3 in the axilla, 2 in the
mandibular region, and one each in the elbow, forearm, back and face.
The histological findings are summarized in Table 5.3.   Among them,
hyperpigmentation of the epidermis, mild hyperkeratosis, severe keratotic
plugs in hair follicles, and moderate chronic inflammatory infiltrate in the
dermis, with foreign body granulomas resulting from the rupture of hair
follicles, were most noticeable (Fig. 5.13).   The changes observed in the
sebaceous glands were rather variable.   In one patient, a 12-yr-old female, the axillary biopsy sample taken 5 months after the onset of skin

TABLE 5.3
Histological Findings for Skin Biopsy of Yusho Patients

Site	No. of samples	Histological findings	Severity				
			‖	‖	+	±	−
Epidermis	24	Hyperkeratosis	0	0	10	14	0
		Pigmentation	0	22	2	0	0
Hair follicle	22	Keratotic plugs	7	8	6	1	0
		Pigmentation	0	0	5	1	16
Dermis	24	Cell infiltrate	0	5	13	1	5
		Granuloma	0	1	6	0	17
		Giant cells	0	2	6	0	16

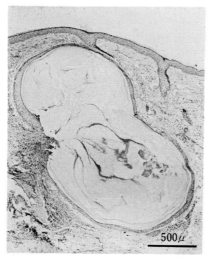

Fig. 5.13 Skin biopsy of 41-yr-old female 2.5 months after the onset of sickness. Cheek. Prominent hyperkeratosis of the hair follicles (H–E stain, ×31).

lesions showed prominent dilatation of and swollen epithelial cells in the sweat glands. Such changes were almost identical with those of the autopsy patient (Case 2) when examined 11 months after the onset of the disease.

5.4 GAS-CHROMATOGRAPHIC ANALYSIS OF THE PCB CONTENT OF VISCERAL ORGANS IN YUSHO PATIENTS[5]

Gas-chromatographic analysis for PCB's was performed on the liver, fatty tissue, heart, kidney and/or brain of the 6 autopsy cases. The extraction and clean up of PCB were made using *n*-hexane and acetonitrile. The extracted fatty element was analyzed by gas chromatography following the method standardized by the Japanese Ministry of Welfare.[13] A quantitative analysis of the PCB was made from the sums of the heights of all peaks which coincided in position with a standard Kanechlor sample, except for the peaks of DDE.

The amounts of PCB in the tissue samples examined are summarized in Table 5.4.   The mean concentration of PCB in the fatty tissue (excluding the stillborn) was 2.6±1.4 ppm on a whole base, and 6.9±4.8 ppm on a fat base.   The controls showed corresponding values of 0.9±0.5 and 1.3±0.9 ppm, respectively.

TABLE 5.4
PCB in Tissues of Autopsy Patients[†]

	Amount of sample (g)	Fat content (%)	PCB (ppm)	
			Fat base	Whole base
CASE 1: 13 yr, M				
Brain	17.7	4.1	0.6	0.02
Fatty tissue	12.5	34.7	3.7	1.3
Liver	10.7	1.5	9.5	0.14
Heart	2.7	3.7	15.0	0.56
Kidney	10.5	1.1	9.6	0.1
CASE 2: 25 yr, M				
Skin	2.5	14.3	8.7	1.2
Fatty tissue	5.4	18.5	15.1	2.8
Liver	12.8	2.0	10.4	0.2
Heart	4.8	28.1	18.3	5.2
CASE 3: 73 yr, M				
Skin	3.0	23.3	4.4	1.0
Fatty tissue	8.2	45.7	8.4	3.8
Liver	6.6	2.3	3.1	0.07
CASE 4: 48 yr, F				
Skin	3.0	79.3	0.8	0.6
Fatty tissue	3.0	70.0	0.9	0.7
Liver	3.0	5.3	1.3	0.07
Heart	3.0	28.3	0.8	0.2
CASE 5: 46 yr, M				
Skin	3.2	56.3	3.2	1.8
Fatty tissue	8.5	67.1	6.5	4.3
Liver	5.5	0.9	8.4	0.08
Heart	1.3	29.6	0.3	0.08
Kidney	6.7	2.0	0.4	0.01
CASE 6: stillborn, M				
Skin	0.5	8.9	1.2	0.01
Fatty tissue	3.0	16.3	0.1	0.02
Liver	8.8	4.0	1.8	0.07
Control (11 cases)				
Fatty tissue			1.3±0.87	0.9±0.46

† PCB values were calculated as Kanechlor 500+600 (1 : 1).

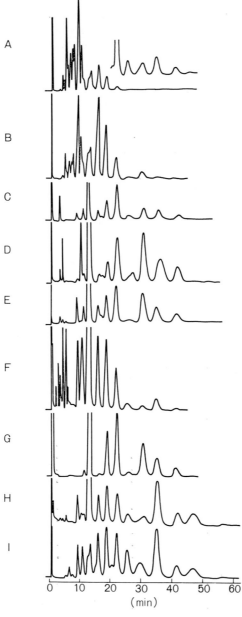

Fig. 5.14 Gas chromatographic patterns for PCB's. A, Kanechlor 400.
B, contaminated rice oil. C, brain (Autopsy Case 1). D, liver (Autopsy Case 2). E, mesenterial fatty tissue (Autopsy Case 3). F, mesen-

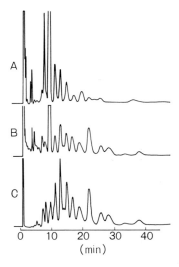

A

B

C

0   10   20   30   40
(min)

Fig. 5.15   Gas chromatograms for PCB's in stillborns.   A, liver (Control stillborn).   B, liver (Autopsy Case 6).   C, Kanechlor 500+600 (1 : 1).

Examples of the gas-chromatographic patterns obtained are given in Fig. 5.14.   The PCB ingested by the Yusho patients contained Kanechlor 400 plus smaller amounts of components with low boiling point. The PCB present in the body about 1 yr after the onset of symptoms indicated that ingredients of high boiling point, corresponding to those of Kanechlor 500 or 600, were retained.   The findings suggested that the patients had ingested large amounts of PCB's mainly with 4 chlorine atoms per molecule, but that most was discharged within 1 yr, leaving only PCB's with 5 to 7 chlorine atoms per molecule in the body.

The gas-chromatographic patterns for PCB's showed almost the same form in all examined organs of the same patient.   The same special patterns, with a remarkably higher peak at 31 min and a lower peak at 16 min compared to the control, were seen in 4 of the 6 cases.   The gas-chromatographic pattern of the stillborn (Fig. 5.15) was almost identical with that of the rice oil which his mother had used.   This may have resulted from the rather short interval between the mother's use of the oil and the death of the child.   Patient 4 had only mild clinical manifestations of Yusho and the gas-chromatographic patters of her organs showed no definitive difference from those of the control.

terial fatty tissue (Autopsy Case 4).   G, liver (Autopsy Case 5).   H, Fatty tissue (control person).   I, Kanechlor 500+600 (1 : 1).   Instrument : Shimadzu GC–5A with ECD[63]Ni.   Column : 5% SE–30 on Chromosorb W AWDMCS.   Column temperature : 200°C.   Note that the curve at the upper right hand of A is 10 times magnified.

5.5 DISCUSSION

The special, consistent findings of Yusho patients considered to be due to PCB's are shown only in the skin and its appendages, viz. as acneform eruptions. Dermatoses of the acneform type have been repeatedly reported in Europe and America as industrial "chloracne",[14] a term which was first proposed by Herxheimer[15] in 1899 to describe the eruptions composed of comedones and small sebaceous pustules which occurred on the arms and faces of workers manufacturing chlorine gas. Thereafter many reports on the occurrence of chloracne have appeared, and chlornaphthalenes, chlorobiphenyls, chlorobiphenyloxides, solid chlorphenols, etc. have been considered as causative substances. Schwertz described histopathologic findings for chlorobiphenyl and chlornaphthalene acne:[16] the characteristic changes were sebaceous cysts rather than comedones. The sebaceous glands were enlarged, some forming huge cysts filled with keratinous material, but with very little sebaceous material. Associated with such cysts were widened follicular openings with keratinous material, and hyperplastic follicular walls. Foreign body granulomas probably due to rupture of the cyst wall and contact of the sebum with the collagen were also present. Jones et al.[17] have described the case of a 26-yr-old negro who worked on the distillation of chlorinated biphenyls; blackheads appeared on his face, neck and other parts of the body surface. Microscopically, the skin lesions chiefly showed cystic dilatation, destruction of the hair, marked thinning and atrophy of the epithelium of the follicles, and a heavy plug of keratinized material which partly filled the cystic cavity. Slight edematous changes were found in some of the sebaceous glands.

Both the biopsy findings on Yusho patients reported here and the skin changes in autopsy patients indicated atrophic epidermis, hyperkeratosis and cystic dilatation of hair follicles, keratinization of the duct epithelium of sebaceous glands, and heavy deposits of melanin in the basal layers. One of the autopsy patients showed prominent atrophy of the sebaceous glands and swelling of the duct epithelium of the sweat glands. Changes in the sweat glands were also observed in one biopsy sample taken from a severe patient. Abnormal perspiration has been reported in many Yusho patients,[12] and so the changes in the sweat glands may perhaps be related to the effects of PCB.

The cutaneous changes of Yusho closely resemble those of occupational acne, although the method of intake of the poisonous substance is different. Whether they arrive directly from the skin surface or through the digestive tract, chlorinated biphenyls thus appear to induce similar cutaneous reactions.

Consistent visceral changes were not detected in the patients, perhaps due to the small number of autopsies.   On the other hand, Herzberg[18] has reported that 7 patients of chlorparaffin poisoning induced by the oral intake of contaminated potatoes experienced diarrhea, abdominal pain and vomiting.   Good et al.[19] have summarized the skin changes experienced by 52 patients of chlorobiphenyl poisoning; however, no patient showed visceral changes.   Bérard[20] has reported repeated pleuritis within 8 months after the onset of chloracne, a condition which he considered might be related to the same toxic substance as induced the chloracne.

Liver disturbacnes are the most noticeable changes accompanying poisoning by chlorine compounds.   Cotter[21] has reported 2 autopsy cases of acute pentachlornaphthalene poisoning, both of whom suffered from severe jaundice.   Histologically, the liver showed a complete loss of liver cells, hemorrhage of the central zone of the lobules and prominent proliferation of the bile duct at its periphery, although inflammatory reactions were found to be mild.   Acute yellow liver atrophy and hepatic necrosis caused by chlornaphthalene and chlorphenol have also been reported by Flinn et al.,[22] Drinker et al.,[23] Greenberg et al.[24] and Dwyer.[25]   Koller et al.[8] have described the influence of polychlorinated biphenyls in rabbits. Mid-zone necrosis of the liver and destruction of the rough-surfaced endoplasmic reticulum of hepatocytes were the most prominent features.   Hirayama et al.[26] have also observed a reduction of the rough-surfaced endoplasmic reticulum, hypertrophy of the smooth-surfaced endoplasmic reticulum, and increased microbodies and lysosomes in biopsy specimens from a Yusho patient (for details, see Chapter 6).   Nishizumi[7] has described an increase in the smooth-surfaced endoplasmic reticulum and lysosomes of liver cells after long-term oral administration of Kanechlor to the mouse and monkey, changes which he ascribed to an adaptation for enhanced drug detoxication.

As described above, on autopsy, one case of liver cirrhosis with hepatic cell carcinoma (Case 4) and one case of amyloidosis (Case 3) were found among the patients examined; however, the other 4 patients showed no remarkable liver changes.   Electron-microscopic examinations of formalin-fixed liver samples in Cases 2 (Fig. 5.16) and 5 revealed increased endoplasmic reticulum and enlarged mitochondria.   The appearance of giant mitochondria in Yusho patients has previously been described by Yamamoto et al.[27] for biopsy material.

In addition to liver lesions, McCune et al.[10] and Flick et al.[9] have demonstrated chlorobiphenyl-induced hydropericardium, hemorrhage of the visceral organs, hypertrophy of the heart, and renal disturbance in chickens.   Koller et al.[8] have reported uterine atrophy after the oral administration of chlorobiphenyls ın rabbits. Goto et al. have induced hydropericardium in chickens by administration of contaminated rice oil.   In the present autopsies, 2 patients showed pericarditis, although in one patient (Case 1) its cause was uncertain.   Severe, special myocardialle-

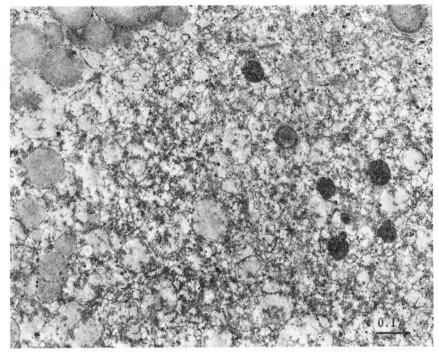

Fig. 5.16 Autopsy Case 2. Liver. Increased microbodies, enlarged mitochondria, reduction of rough-surfaced endoplasmic reticulum, and hyperplasia of smooth-surfaced endoplasmic reticulum (×10,000).

sions with small fibrotic and necrotic foci and basophilic myofibrillar degeneration, together with a large amount of PCB in the heart, were observed in this patient. The possibility that PCB's may have a strong effect on the heart and pericardium thus remains. McCune et al.[10] have reported the intramyocardial infiltration of lymphocytes and heterophils. Reichenbach et al.[28] have described similar basophilic fibrillar degeneration of myocardial fibers in cases of subarachnoidal hemorrhage, pheochromocytoma, and after cardiac operations. They explained the lesions as possibly related to a disturbance of catecholamine or isoproterenol metabolism. Certain Yusho patients are known to have experienced metabolic disturbance of the endocrine system,[29] indicating that the myocardial lesions in Case 2 might perhaps be related to hormonal disturbance. However, the possibility of viral infection or toxic drug action cannot be ruled out. The pericarditis observed in Case 3 was probably associated with the amyloid deposition affecting the heart.

Changes in the esophageal glands were noted in 3 of the 6 cases. Lesions of the esophagus have been reported by Takeuchi et al.,[30] who described hyperkeratosis of the squamous epithelium of mice. The affected patients showed 4 or 5 layers of hypertrophic epithelial cells in the ducts and the changes resembled those in the duct of the sweat glands of Case 2. The cause of the lesions is uncertain, but the resemblance to changes in the sweat glands and the oral intake of Kanechlor are suggestive of a PCB influence. Other reported visceral lesions due to PCB's include tubular changes of the kidney and enteritis in chickens,[9],[10] fatty degeneration and swelling of the proximal renal tubules in rats.[30] However, no such lesions were detected in the present autopsy cases. Case 5 exhibited the features of osteitis fibrosa generalisata with disseminated metastatic calcification, especially in the heart, lungs and liver. The cause of disturbance of calcium metabolism in this patient was not determined, although clarification of the nature of the lesion and its possible relation to PCB intake undoubtedly represents an important question.

The histological changes in the stillborn showed differences from those in other cases. Severe darkish pigmentation of the skin was the prominent feature, other changes being relatively mild. Clinically, 9 such darkish brown babies born to Yusho mothers have been reported by Taki *et al.*[31] The differences in the lessions from other Yusho patients probably arose from the age difference and method of intake. Thus, from the viewpoint of PCB-induced changes after oral intake, 2 types, adult and fetal, may be distinguished.

Examinations of the distribution of Kanechlor in rats revealed an accumulation of PCB's related to the fatty tissues. In the present autopsy cases, the highest concentrations of PCB were found in the mesenterial fatty tissue, followed by the skin, liver and heart, whereas the PCB concentration in the brain was low. The PCB concentrations and clinical manifestations appeared to be related in the adult type; however, in the fetal type, in spite of the low concentrations, the cutaneous pigmentation was severe. The concentration rates of PCB in the autopsy cases displayed no definitive differences from those in control cases, suggesting that the cutaneous changes in the autopsy patients might be residual ones occurring at the acute or subacute toxic stage. However, the special pattern observed in the gas-chromographs, unexplained cardiac changes, and disturbance of calcium metabolism could indicate a late action for PCB's. It is thus necessary to continue work on the progress of Yusho patients.

REFERENCES

1. H. Tsukamoto *et al.*, *Fukuoka Acta Med.*, **60**, 496 (1969).
2. M. Kuratsune, *Environ. Health Persp.*, **1**, 129 (1972).
3. M. Kikuchi, Y. Mikagi, M. Hashimoto and T. Kojima, *Fukuoka Acta Med.*, **62**, 89 (1971).

4. M. Kikuchi, *ibid.*, **63**, 41 (1972).
5. M. Kikuchi and Y. Masuda, *Jap. J. Clin. Path.*, **21**, 5 (1973).
6. M. Kikuchi, M. Hashimoto, M. Hozumi, K. Koga, S. Oyoshi and M. Nagakawa, *Fukuoka Acta Med.*, **60**, 489 (1969).
7. M. Nishizumi, *Arch. Environ. Health*, **21**, 620 (1970).
8. L. Koller and J. G. Zinkel, *Am. J. Path.*, **70**, 363 (1973).
9. D. F. Flick, R. G. O'Dell and V. A. Childs, *Poultry Sci.*, **44**, 1460 (1965).
10. E. L. McCune, J. E. Savage and B. L. O'Dell, *ibid.*, **41**, 295 (1962).
11. M. Kikuchi and M. Hashimoto, *Fukuoka Acta Med.*, **60**, 484 (1969).
12. M. Goto and K. Higuchi, *ibid.*, **60**, 409 (1969).
13. Y. Masuda, R. Kagawa and M. Kuratsune, *ibid.*, **65**, 17 (1974).
14. E. W. Baader and H. J. Bauer, *Ind. Med. Surg.*, **20**, 286 (1951).
15. K. Herxheimer, *Munch. Med. Wschr.*, **46**, 278 (1899).
16. J. Schwartz, *Occupational Diseases of the Skin*, 3rd ed., p. 336, Lea and Febiger, 1957.
17. J. W. Jones and H. S. Alden, *Arch. Derm. Syph.*, **33**, 1022 (1936).
18. H. J. Herzberg, *Derm. Wschr.*, **7**, 425 (1947).
19. C. K. Good and N. Pensky, *Arch. Derm. Syph.*, **48**, 251 (1943).
20. J. Bérard, *J. Franc. Med. Chir. Thorac.*, **19**, 87 (1965).
21. L. H. Cotter, *J.A.M.A.*, **125**, 273 (1944).
22. F. B. Flinn and D. E. Jarvik, *Proc. Soc. Exptl. Biol. Med.*, **35**, 118 (1936).
23. C. K. Drinker, M. F. Warren and G. A. Bennett, *J. Ind. Hyg. Toxicol.*, **19**, 283 (1937).
24. L. Greenburg, M. R. Mayers and A. R. Smith, *ibid.*, **21**, 29 (1939).
25. J. M. Dwyer, *Med. J. Australia*, **2**, 703 (1946).
26. C. Hirayama, T. Irisa and T. Yamamoto, *Fukuoka Acta Med.*, **60**, 455 (1969).
27. T. Yamamoto, C. Hirayama and T. Irisa, *ibid.*, **62**, 85 (1971).
28. D. D. Reichenbach and E. P. Benditt, *Human Path.*, **1**, 125 (1970).
29. A. Watanabe, S. Irie, T. Nakajima and S. Katsuki, *Fukuoka Acta Med.*, **62**, 159 (1971).
30. T. Takeuchi, N. Tomio, K. Eto, H. Matsumoto, A. Fujisaki, M. Kodama and S. Suko, *Kumamoto Acta Med.*, **43**, 63 (1969).
31. I. Taki, S. Hisanaga and Y. Amagase, *Fukuoka Acta Med.*, **60**, 471 (1969).

6

Clinical Aspects of PCB Poisoning

Chisato HIRAYAMA

6.1 INTRODUCTION

Despite some recent studies, the toxicology of polychlorinated biphe-nyls (PCB's) remains rather poorly known in comparison with that of chlorinated hydrocarbon pesticides such as DDT and BHC. Typically, PCB's appear to have two distinct actions on the body, viz. a skin effect and a liver effect;[1] however, details were not known until 1968, when poisoning by PCB's occurred accidentally in Japan.

In 1968, over 1000 persons in southwest Japan were affected by PCB poisoning (Yusho) after peroral ingestion of rice oil contaminated with Kanechlor 400, the main component of which was tetrachlorobiphenyl. The average amount of PCB's ingested has been estimated at about 1.0 g per person. The major clinical symptoms have been characterized as acneform eruptions,[2] peripheral neuropathy,[3] hepatomegaly with no disturbance of routine liver tests,[4,5] irregular menstruation in women,[6] and growth retardation in infants and children.[7,8] It seems worthwhile

87

therefore to summarize and evaluate our present knowledge of PCB poisoning in humans.   The dermatological features of Yusho are dealt with in detail in Chapter 7.

6.2   CLINICAL MANIFESTATIONS

Like chlorinated hydrocarbon pesticides, PCB's cause typical acneform eruptions, but they are not toxic to the central nervous system.[9] According to our observations, the initial symptoms of PCB poisoning are not specific, and include general fatigue, loss of appetite, nausea, vomiting, and swelling of the limbs.   Specific symptoms, such as increased discharge from the eyes, swelling of the eyelids, acneform eruptions, pigmentation of the nails, gingivae and lips, may follow.[2,10]   Since PCB's are soluble in oils, those absorbed from the intestine are distributed and concentrated preferentially in adipose tissues or tissues with a high lipid content.   It seems therefore that the symptoms of PCB poisoning may depend on either the lipid content and/or susceptibility of the target organs.   The non-dermal manifestations of PCB poisoning may be classified and discussed under the following headings: disorders of the nervous system, endocrine disorders, respiratory disorders, hematologic disorders, hepatic disorders, metabolic disorders, bone and joint disorders,[11] and effects on fetal and infant life. (Table 6.1).

TABLE 6.1
Major Manifestations of PCB Poisoning

1.	Neurologic disorders ····················	Disorders of the central nervous system and peripheral nerves
2.	Endocrine disorders ····················	Adrenocortical and gonadal dysfunction
3.	Respiratory disorders ·················	Chronic bronchitis
4.	Hematologic disorders ················	Immunologic deficiency
5.	Hepatic disorders ····················	Hepatomegaly with enzyme induction
6.	Metabolic disorders ····················	Hypertriglyceridemia
7.	Bone and joint disorders·············	Metabolic bone disease(?)
8.	Dental disorders·························	Developmental abnormalities
9.	Ophthalmic disorders ·················	Meibomian gland disorders
10.	Dermatologic disorders···············	Acneform eruptions and abnormal pigmentation
11.	Effects on infants and children ······	Growth retardation

6.2.1   Neurologic disorders

About half of the Yusho patients complained of headaches, numbness, hypoesthesia, and neuralgic limbs, features which resemble DDT

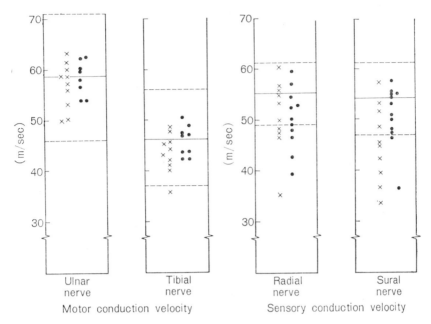

Fig. 6.1 Nerve conduction velocities (m/sec) in patients with PCB poisoning.[3] Broken lines indicate normal ranges; crosses and dots indicate patients with and without neuropathic symptoms, respectively.

poisoning.   To elucidate such manifestations, neurological and electrophysiological examinations were carried out.   Most of the headaches were transient, but some were chronic and recurrent over periods of months or years.   However, no symptoms or signs were found which indicated involvement of the cerebellum, spinal cord or cranial nerves.   In fact, electroencephalographic examinations of 20 patients resulted in only one case with a slightly abnormal record of the low voltage, 6–7 c/s theta wave of the frontal dominant.[12]   These findings suggested that most of the headaches arose from a migraine condition or emotional stress.

Further careful neurological examinations revealed certain spinal nerve symptoms and signs.   To elucidate the peripheral nerve involvement in the patients, the sensory and motor nerve conduction velocities were examined[3] (Fig. 6.1).   In almost all patients, the motor nerve conduction velocities in the ulnar and tibial nerves were found to be within the normal range.   On the other hand, there was slowing of the sensory nerve conduction velocity in the radial and sural nerves in 8 (38%) and 7 (33%) cases, respectively.   In 10 cases (48%), the sensory nerve conduction velocity was lower than normal in the radial or sural nerve or both.   In the neuropathic group, 6 cases (75%) showed slowing of the conduction velocity in the sural nerve, whereas only 1 case (8%) in the non-neuropathic group showed such a reduced conduction velocity.

Acute feeding experiments of PCB's to animals revealed that PCB's are 1/4 to 1/5 as toxic as DDT for the central nervous system.[9]   Also, post-mortem examinations of several patients indicated a PCB concentration in the brain of only 1/40 of that in adipose tissues.[13]   Neurological manifestations of PCB poisoning were thus not expected to be very marked.   Indeed, present clinical observations suggest that, unlike DDT poisoning PCB poisoning does not include any significant involvement of the central nervous system, but rather of the peripheral nerves, almost exclusively sensory nerves.   In cases of unspecified polyneuropathies, however, motor conduction may be affected after sensory conduction.[14] It seems probable therefore that the present neurological and electrophysiological findings of patients may have revealed only the early manifestations of wider neurological effects.   To elucidate the full impact of PCB's on the nervous system, further follow-up studies are thus required.

6.2.2   Endocrine disorders

In the early stages of PCB poisoning, certain patients are known to have died suddenly.   The symptoms and signs of these patients resemble those of adrenal crisis, e.g. profound asthenia, severe pain in the abdomen, nausea, vomiting, arterial hypotension associated with peripheral vascular collapse, etc.   The crisis is precipitated most often by hard muscular work, presumably through salt loss with excessive sweating.   Post-mortem examinations did not reveal any consistent findings related to adrenocortical morphology.   However, the concentration of PCB's in the adrenals proved to be about 1/3 of that in common adipose tissues.[13]   It thus appeared likely that PCB's might interfere with certain endocrine functions.   Between 1969 and 1970, viz. 1–2 yr after the onset of PCB poisoning, clinical observations on endocrine functions were performed on Yusho patients.

Fasting serum levels of 11 OHCS were in the normal range for 72% of the patients, and high in the remainder.[15]   Rapid ACTH tests showed no evidence of severe abnormalities in adrenocortical function, except 2 cases with a low response (Fig. 6.2).   Urinary excretions of 17 OHCS and 17 KS were elevated in about 40% of the patients.[16]   The major components of urinary 17 KS, e.g. androsterone, etiocholanolone and dehydroepiandrosterone, were found to increase in male patients, but to decrease in female patients.   These findings thus did not lead to any definite conclusions concerning adrenocortical hypofunction, but did suggest some degree of adrenocortical dysfunction in the patients.   It is reasonable to assume that in the early stages of PCB poisoning, certain stresses may cause adrenal insufficiency, followed by sudden death.

In female patients, symptoms suggesting abnormal ovarian function also appeared.[6]   Over 60% of the patients showed abnormalities in the

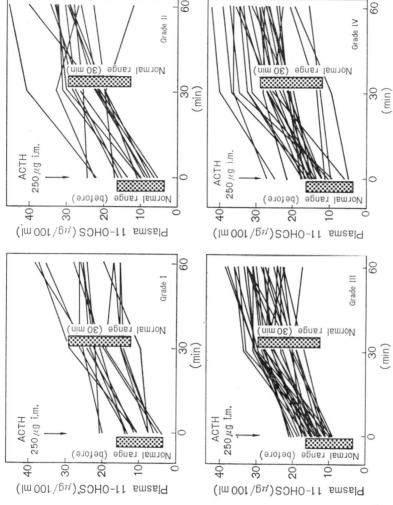

Fig. 6.2 Rapid ACTH tests in patients with PCB poisoning,[18] Grades I—IV indicate increasing grades of severity of dermal symptoms (see Chapter 7, sect. 2.12).

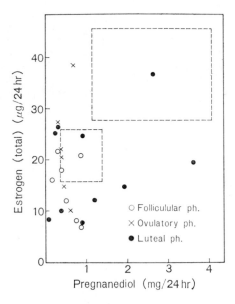

Fig. 6.3 Relationship between urinary excretions of estrogen and preg-nanediol in patients with PCB poisoning.[6]

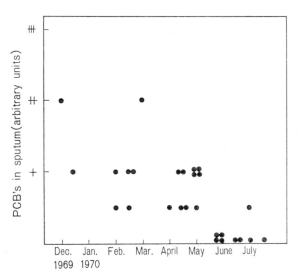

Fig. 6.4 PCB's in the sputum of patients with PCB poisoning (as detected before July, 1970).[20]

menstrual cycle, the menstrual interval being either prolonged, shortened or irregular.   In about 55% of the patients, abnormal durations of menstrual flow were observed, the quantity of menstrual flow being increased and/or diminished.   The basal body temperature showed an abnormal pattern in 85% of the patients (i.e. a poor high phase and a prolonged low phase), suggesting the existence of corpus luteum insufficiency, retarded follicular maturation, or no ovular cycle.   Serial determinations of urinary estrogens revealed a low output in the patients.   Urinary excretion of pregnanediol was also low, especially in the luteal phase.   The relationship between the urinary excretions of estrogen and pregnanediol is illustrated in Fig. 6.3.   Despite the wide range of normal values, urinary estrogen and pregnanediol tended to decrease, especially in the luteal phase.   These abnormalities may have resulted from either primary ovarian disorders, secondary suppression by other glands, or some abnormal function of the target organs.   Abnormal degradation of sex hormones by the liver should also be considered as a possible factor, as discussed later.

6.2.3   Respiratory and hematologic disorders

About 40% of the patients complained of coughing with sputum production.   The physical signs and symptoms, X-rays, and bronchographic findings all suggested the existence of chronic bronchitis.[17]   As in common chronic bronchitis, most of the patients were afebrile, with little change in leucocyte count or erythrocyte sedimentation rate.   Respiratory functions were also within normal limits, although some patients showed mild small-airway obstruction.[18] Non-specific respiratory infections, dominantly of *Candida*, *Staphylococcus* and α-*Streptococcus*, were often detected. This condition may have involved particularly those patients who had a previous history of respiratory disease.   PCB poisoning, therefore, may be causally related to the chronic bronchitis.

Following the peroral ingestion of PCB's in oil, most PCB's enter the blood stream via the thoracic duct, and so reach the respiratory system. Since pulmonary tissues metabolize lipids very actively,[19] PCB's appear to become distributed throughout the lungs and bronchi.   In fact, PCB's have been detected in the lung and sputum of patients at least before May 1970[20]   (Fig. 6.4).   The PCB's are then presumably excreted via the respiratory tract, and simultaneously cause respiratory distress.   The exact cause of the chronic bronchitis still remains obscure, although PCB's *per se* appear to affect some defense mechanism.   Experimental studies have indicated that PCB's cause a decrease in the acid phosphatase activity of alveolar macrophages.[18]

Hematologic examinations revealed slight leukocytosis and monocytosis, and the bone marrow picture showed elevated plasmacytosis and reticulocytosis.[21]   These findings may correspond to the hematologic features of inflammatory disorders.   The serum levels of immunoglobulins, especially IgA and IgM, were generally decreased,[17] suggesting B cell dysfunction.   Although T cell function is not known, PCB's are reported to be capable of immunosuppressive activity.[22]   Therefore, the chronic infections generally observed in the patients may be the result of some immunological deficiency.   At present, however, it is difficult to determine whether the PCB's effect on defense mechanisms, including the immune mechanism, is local or general.   For example, hepatitis B antigen was detected in only 2.6% of the serum samples from the patients,[23] indicating no significant difference in incidence between the patients and healthy blood donors.

6.2.4 Hepatic disorders

Initial biochemical findings on the patients have revealed no abnormalities in liver tests, such as serum bilirubin, GOT, GPT and BSP retention, apart from a slightly elevated level of alkaline phosphatase.[4]   Serum protein electrophoresis showed slightly decreased albumin and elevated α_2 globulin, as observed in acute non-specific poisoning. Some patients, however, did have a slightly enlarged liver.

Studies on hepatic biopsy specimens from one patient revealed no changes by light microscopy, but definite changes were evident in electron micrographs.[5,24]   The rough-surfaced endoplasmic reticulum was reduced, whereas the smooth-surfaced endoplasmic reticulum showed a marked proliferation (Fig. 6.5).   The mitochondria exhibited morphological heterogeneity, i.e. variations in size and form.   Inclusion bodies were recognized both in the matrix and within the cavity of cristae.   Lysosomes and microbodies were present in increased numbers.   The microbodies, located close to the smooth-surfaced endoplasmic reticulum, were increased in size and contained no crystalline structures.   These findings were initially interpreted as partially adaptive, rather than as degenerative phenomena, but subsequently were considered as typical morphological features closely related to the enzyme induction caused by PCB's.   In fact, the possible hepatic changes so far observed to be induced by such pollutants in monkeys are hepatomegaly with no disturbance of routine liver tests,[25] as confirmed in this study.

Like DDT, PCB's are known to be powerful inducers of hepatic enzyme systems, e.g. microsomal drug-metabolizing enzymes and mitochondrial δ-aminolevulinate synthetase.[7,26,27]   The mechanisms concerned in microsomal enzyme induction involve an increased rate of metabolism of endogenous substances, such as hormones, and presumably bilirubin.   For example, PCB's cause increased degradation of hormones, including estradiol.   Such diminished circulating estradiol may thus be responsible

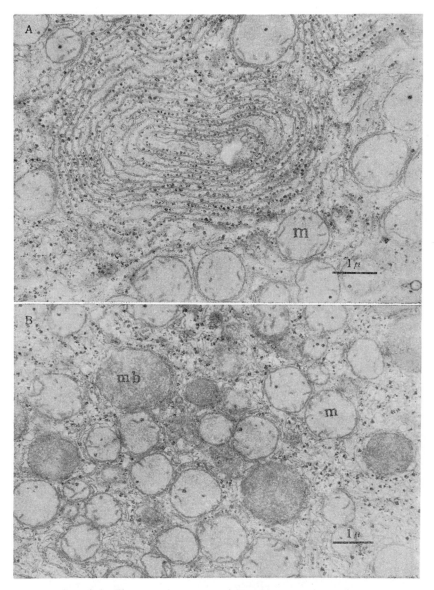

Fig. 6.5 Electron micrographs of liver biopsy specimens from a patient with PCB poisoning.[5,24] The smooth-surfaced endoplasmic reticulum is increased and the rough-surfaced endoplasmic reticulum reduced (A); Mitochondria show variation in size and form, and microbodies are enlarged (B).

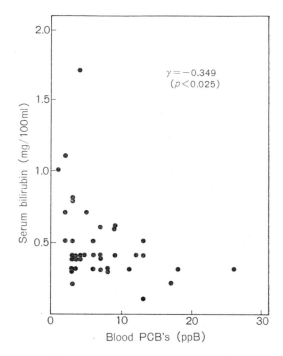

$\gamma = -0.349$
$(p < 0.025)$

Fig. 6.6   Relationship between serum bilirubin concentration and blood PCB concentration in 40 patients with PCB poisoning.[28]

for the menstrual disorders observed in female patients.   In addition, a lowered concentration of serum bilirubin has been observed in patients.[28] The relationship between the serum levels of bilirubin and PCB's is illustrated in Fig. 6.6   It indicates a significant inverse correlation.   The effect of enzyme induction on bilirubin metabolism is thus regarded as either an augmentation of bilirubin conjugation and/or an enhancement of hepatic excretion, resulting in hypobilirubinemia.   Experimental data have also shown that PCB's may cause hepatic porphyria in the Japanese quail through the induction of δ-aminolevulinate synthetase in the mitochondria.[27]   However, at present no clinical symptoms suggestive of hepatic porphyria have been observed in human patients.

Experimental work on animals has shown that PCB's are 40 to 300 times less toxic than DDT, the toxicity decreasing with increase in the chlorine content.[9]   On the other hand, the chlorine content of PCB's shows a positive correlation with the degree of enzyme induction.   Although Kanechlor 400 is composed mainly of tetrachlorobiphenyl, trace amounts of penta- and hexachlorobiphenyls, which are difficult to excrete, continuously stimulate hepatic enzyme systems.   Although the clinical manifestations responsible for hepatic enzyme induction are not yet satisfactorily elucidated, it seems likely that they involve endocrine disorders. The complete ineffectiveness of drug therapy observed in some patients may appear either through a low response of the target organ(s) or an increased degradation of the administered drugs by the liver.

6.2.5 Metabolic disorders

As discussed above, metabolic changes in the patients are very characteristic. These alterations appear to be caused primarily by the dysfunction of metabolic organs, and secondarily by accelerated metabolism through enzyme induction. Since ingested PCB's are concentrated in adipose tissues or the skin, distinct dermatological manifestations appear. For example, the most characteristic acneform eruptions may be based on hyperkeratization associated with abnormal lipid metabolism in the skin. In fact, alterations in general lipid metabolism are probably the most important metabolic change to occur in the patients. Serum lipid analysis has revealed an elevated concentration of triglyceride (Fig. 6.7) but nor-

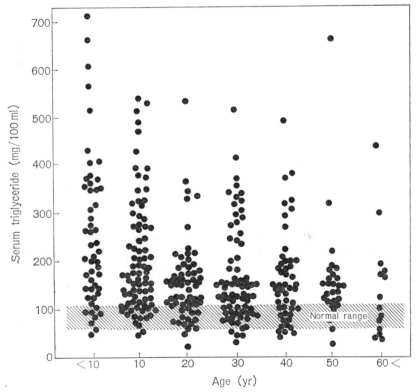

Fig. 6.7 Serum triglyceride levels in patients with PCB poisoning (1970).[31]

mal levels of cholesterol and phospholipid.[29]   Since a significant positive correlation between the serum triglyceride level and the blood PCB concentration was observed,[30] hypertriglyceridemia appears to be caused by PCB's *per se*.   Agarose gel electrophoresis of serum lipoproteins showed faint α, dense pre-β, no tailing behind β, and no chylomicron at the origin. These results indicate that the elevated triglyceride is of endogenous origin.[29]

Hypertriglyceridemia, a characteristic feature of PCB poisoning, can be caused by either increased synthesis or a decreased removal of triglyceride from the blood. Since hyperfunction of the adrenal cortex could perhaps occur in the early stages of poisoning, excess free fatty acids might be released from the adipose tissue and subsequently accumulate in the liver, where they would be synthesized into triglyceride. Because there is no abnormality in hepatic function, excess triglyceride would then pass into the blood, causing hypertriglyceridemia. However, since hypertriglyceridemia is still observed at present, several years after poisoning, another process may be implicated. It thus seems more likely that there is disturbance of triglyceride removal in the patients. Most of the serum triglyceride is hydrolyzed and incorporated into adipose tissue, and it is well known that there is a significant correlation between the ability of adipose tissue to incorporate triglyceride fatty acids and the activity of the enzyme lipoprotein lipase, which is closely related to postheparin plasma lipoprotein lipase. In the patients, especially female patients, postheparin lipoprotein lipase activity was actually decreased.[31] The hypertriglyceridemia, therefore, appears to be caused by a disturbance of plasma triglyceride removal through diminished lipoprotein lipase, which is a result of the direct effect of PCB's on adipose tissue.

Besides triglyceride, there are some findings which suggest abnormal calcium metabolism.[11] Though serum levels of calcium and phosphorus were found to be normal,[4] some patients claimed bone and dental disorders, such as osteomalacia, osteoporosis, dental caries and periodontal disease, although these have not yet been confirmed. On the other hand, post-mortem findings in one case, who died through acute heart failure with pneumonia, have revealed calcium deposition in several organs including the heart.[32] Clinical symptoms and signs did not support the existence of hyperparathyroidism. Accordingly, if calcium metabolism is actually disturbed in the patients, some other mechanism may be involved. As discussed above, the occurrence of enzyme induction tends to diminish certain endogenous and exogenous substances, such as sex hormones and vitamins. Normally, vitamin D_3 (cholecalciferol) is transported to the liver, where it is converted into 25-hydroxycholecalciferol by a mitochondrial enzyme system which requires NADH and molecular oxygen. This transformation system seems to be strongly enhanced by PCB's, as observed in microsomal enzyme systems. Thus, PCB's may induce some estrogen and/or vitamin D deficiency.

It is well known in birds that the accumulation of a high dose of PCB's causes disruption of the normal breeding behavior and the formation of thin-shelled eggs.[26]   Calcium metabolism in birds is intimately related to reproductive metabolism and is to a large extent regulated by steroids, such as estrogen and vitamin D.   The deposition of medullary bone, the chief source of calcium during egg and egg-shell formation, is controlled by the steroid sex hormones, and hens deficient in vitamin D lay eggs with lower egg-shell weights.   Based on these facts, a further precise clinical assessment of calcium metabolism in Yusho patients is required.

6.3   Effects on Fetal and Infant Life

Pregnant women who suffered from PCB poisoning delivered babies (occasionally stillborn) with abnormal characteristics.[7,33] PCB's can thus be said to affect fetal life, probably through the placenta.   In fact, most newborn babies were in the weight range of small-for-date babies (Fig. 6.8).   Further studies also revealed that neonates fed mainly on mother's milk (i.e. where the mother had PCB poisoning) showed characteristic symptoms of PCB poisoning, indicating that PCB's may reach the newborn through the mother's milk.[34]   Although laboratory examinations of affected babies have shown few biochemical abnormalities, typical dark-brown pigmentation and parchment-like desquamation of the skin

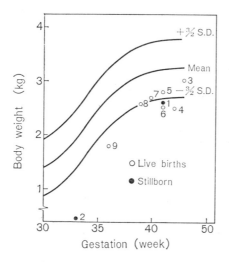

Fig. 6.8 Body weights of fetuses and infants of women with PCB poisoning.[7]   The normal range is assumed to be the mean ±3/2 standard deviations (S. D.).

have been recorded.   Moreover, developmental tooth and bone defects have been noticed, i.e. eruption of teeth at birth, larger frontal and occipital fontanelles than usual, and maintenance of a wider sagittal suture than usual.[35]   Spotted calcification was also noticed on the skull.   These findings strongly suggest a deviation from normal calcium metabolism. During the first 2–5 months after delivery, the pigmentation gradually faded, and the pattern of postnatal growth of these infants paralleled that of healthy infants as regards physical and mental development.

The clinical manifestations of PCB poisoning in infants and children resemble those of adult patients, except for the growth retardation.   Controlled studies have been carried out on the growth of school-children.[8] In 1967, 1968 and 1969, affected children (23 boys and 19 girls) were compared with their matched controls for body height and weight.   The gain in height or wieght before PCB poisoning from 1967 to 1968, and thereafter from 1968 to 1969, was compared against the distribution pattern of the corresponding gains made in the control group.   Using a fine statistical method, the results obtained showed that both the body height and weight gain of affected children (especially the boys) were significantly decreased after poisoning, while follow-up surveys indicated some growth recovery in the children 2–3 yr after poisoning.   These findings thus suggest that PCB poisoning causes growth retardation, at least when the PCB content in the body remains above a certain level.   At present, however, it is still obscure whether or not growth retardation is caused primarily by PCB poisoning. Further studies concerning endocrine systems, especially observations on growth hormones, are required.

6.4   LABORATORY FINDINGS

PCB poisoning may be confirmed by its characteristic dermatological manifestations, as discussed elsewhere.   Routine laboratory tests are non-specific, but hypertriglyceridemia without hypercholesterolemia may also be utilized as a diagnostic criterion, since serum levels of triglyceride were still found to be elevated in 1972 (mean, 133 ± 64 mg/100 ml).   For confirmation of a state of enzyme induction, decreases in serum sex hormones and serum bilirubin may be useful, although observations on the plasma half-life of certain drugs, and the urinary excretion of glucaric acid, may represent more reliable methods for confirming enzyme induction.[36] These tests, however, are not easy to perform clinically.   Moreover, enzyme induction is also caused by the administration of certain drugs, such as sedatives and analgetics, which are used occasionally for gaining symptomatic relief.   Finally, the PCB content of available tissues (including blood and milk) represents an important diagnostic criterion, provided suitable control data are available.

PCB analysis of biological materials is a time-consuming process.   One conventional method is to extract the tissue samples with hexane, pass the cleaned-up extract through a silica gel column, and then analyze for PCB's by gas chromatography using an electron capture detector.[37]   The PCB concentrations of fatty tissues from Yusho patients have been found to average about 2.5 ppm, which is significantly higher than the 0.90 ± 0.46 ppm for healthy controls living in northern Kyushu.

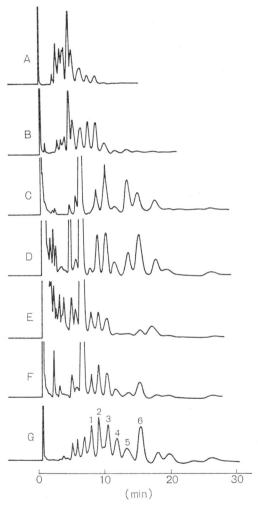

Fig. 6.9 Gas–chromatographic patterns of blood PCB's.[38]   A, Kane-chlor 400; B, rice oil contaminated with Kanechlor 400; C, D and E, blood of patients with PCB poisoning; F, blood of healthy control; G, Kanechlor 500+600 (1 : 1).

Analysis of gas-chromatographic patterns may serve as a more accurate tool, since most tetrachlorobiphenyls (the major components of Kanechlor 400) return to almost normal levels, but penta- and hexachlorobiphenyl levels remain high, a result which is interpreted on the basis of the retention of only metabolically inert components of Kanechlor 400. The behavior of blood PCB's is similar to that of tissue PCB's,[38,39] so that blood PCB levels are assumed to reflect body PCB's, although conclusive evidence of this is still lacking.   The blood levels of PCB's averaged 6.7±5.3 ppb (range, 2 to 28 ppb) in Yusho patients, and 2.8±1.6 ppb (range, 1 to 7 ppb) in healthy controls. Typical gas-chromatographic patterns of blood PCB's are illustrated in Fig. 6.9.   Most patients showed retention of penta- and hexachlorobiphenyls in their blood, so confirming the usefullness of blood PCB analysis for the diagnosis of PCB poisoning, provided suitable control patterns are available.

6.5   TREATMENT

As in the case of poisoning by chlorinated hydrocarbon pesticides,[40] the treatment of acute PCB poisoning is fundamentally symptomatic and depends largely on the stage of poisoning and condition of the patient. The first step is to remove PCB's by emesis or lavage, if they were taken perorally.   After an initial emesis or lavage, the use of a sodium chloride cathertic may be helpful.   However, since oils may promote the absorption of PCB's, the use of oily substances (including laxatives) should be avoided.   To inhibit PCB absorption, certain adsorbent materials such as aluminum silicate and medical carbon could be given.   Cholestyramine, a basic anion-exchange resin, is known to combine with bile salts, which are necessary for oil absorption.   Experimental results have suggested that cholestyramine inhibits PCB absorption, provided the time lag is not longer than several hours after PCB ingestion.[41]

Contrary to our expectations, most PCB's (at least those with 4 or less chlorine atoms per molecule) were rapidly excreted. However, PCB's with 5 or more chlorine atoms are apparently excreted only with difficulty, since they are barely metabolized by the hepatic microsomal systems in which PCB hydroxylation occurs.   Barbiturates, which are known to induce hepatic enzyme systems, were able to accelerate PCB hydroxylation. According to our experience, short-term treatment with barbiturates may cause an increase in urinary glucuronate, although the clinical manifestations could not be eased.[42]   At present, the major excretory channel of PCB's is unknown.   Most hydroxylated PCB's are presumable expelled through the biliary tract into the intestine, so that part of them might be reabsorbed.   Although the adsorbent materials mentioned above are available for inhibition of such PCB reabsorption, clinical tests have not yet been made.

The presence of PCB's in the body fat does not appear to have any serious effect upon body functions, except for the production of skin lesions. However, continuous release of PCB's from adipose tissues may well affect other body functions. If the patients are in sufficiently good condition to receive drastic therapy, starvation may be a useful technique for removing PCB's from adipose tissues. In fact, experimental studies have revealed that the excretion of DDT was accelerated in rats by starvation.[43] Twenty patients complaining of several neurological and dermatological symptoms were therefore subjected to starvation therapy for 8 to 14 days.[44] The method consisted of an absolute fast apart from water for 2 days, and then feeding of fruit or juice (occasionally milk) on subsequent days. During the latter phase, most patients complained of exaggerated nausea and fatigue, but thereafter, and especially some time after treatment, a drastic beneficial effect on neurological symptoms such as heavy headaches, respiratory disorders and nausea was observed in at least 5 of the patients. The exact mechanism responsible for this symptomatic relief remains obscure. To confirm the effectiveness of starvation therapy, further objective studies (especially on PCB excretion) are thus required.

As mentioned above, the treatment of acute PCB poisoning is fundamentally symptomatic. If central nervous stimulation occurs, sodium pentobarbital is given for its quick action, and is followed by phenobarbiturates for more prolonged action. As discussed above, barbiturates also accelerate the excretion of PCB's. Calcium glucuronate may be valuable as a supplement to barbiturates. In severe cases, where adrenal insufficiency is suggested, it may be necessary to administer corticosteroids in appropriate amounts. It is necessary to bear in mind that PCB poisoning tends to cause adrenal crisis in acute distress and during surgical operations on patients. Several symptoms and signs could be controlled by symptomatic therapy. Analgetics and sedatives, which are often administered, tend to be required in larger doses, since they are metabolized more rapidly through the enzyme induction caused by PCB's.

REFERENCES

1. S. N. Irving, *Dangerous Properties of Industrial Materials*, Van Nostrand-Reinhold, 1968.
2. M. Goto and K. Higuchi, *Fukuoka Acta Med.*, **60**, 409 (1969).
3. Y. Murai and Y. Kuroiwa, *Neurology*, **21**, 1173 (1971).
4. M. Okumura and S. Katsuki, *Fukuoka Acta Med.*, **60**, 440 (1969).
5. C. Hirayama, T. Irisa and T. Yamamoto, *ibid.*, **60**, 455 (1969).
6. M. Kusuda, *Obster. Gynecol.*, **38**, 1063 (1971).

7. I. Taki, S. Hisanaga and Y. Amagase, *Fukuoka Acta Med.*, **60**, 471 (1969).
8. T. Yoshimura, *ibid.*, **62**, 109 (1971).
9. D. B. Peakall and J. L. Lincer, *BioSci.*, **20**, 958 (1970).
10. M. Kuratsune, Y. Morikawa, T. Hirohata, M. Nishizumi, S. Kohchi, T. Yoshimura, J. Matsuzaka, A. Yamaguchi, N. Saruta, N. Ishinishi, E. Kunitake, O. Shimono, K. Takigawa, K. Oki, M. Sonoda, T. Ueda and M. Ogata, *Fukuoka Acta Med.*, **60**, 513 (1969).
11. C. Hirayama, *Jap. J. Clin. Med.*, **31**, 2000 (1973).
12. K. Nagamatsu and Y. Kuroiwa, *Fukuoka Acta Med.*, **62**, 157 (1971).
13. M. Kikuchi, Y. Mihagi and M. Hashimoto, *ibid.*, **62**, 89 (1971).
14. W. T. Liberson, *Arch. Phys. Med.*, **44**, 313 (1963).
15. A. Watanabe, S. Irie, T. Nakajima and S. Katsuki, *Fukuoka Acta Med.*, **62**, 159 (1971).
16. J. Nagai, M. Furukawa, A. Jojo and T. Fukimoto, *ibid.*, **62**, 51 (1971).
17. N. Shigematsu, Y. Norimatsu, T. Ishibashi, M. Yoshida, S. Suetsugu, T. Kawatsu, T. Ikeda, R. Saito, S. Ishimaru, T. Shirakusa, M. Kido, K. Emori and H. Toshimitsu, *ibid.*, **62**, 150 (1971).
18. N. Shigematsu, S. Ishimura, T. Hirose, T. Ikeda, K. Emori and N. Miyazaki, *ibid.*, **65**, 88 (1974).
19. H. O. Heinemann and A. P. Fishman, *Physiol. Rev.*, **49**, 1 (1969).
20. T. Kojima, *Fukuoka Acta Med.*, **62**, 25 (1971).
21. M. Kozuru, S. Motomura, K. Sakai, H. Hiraiwa, Y. Hatta, K. Takuchi, K. Inoue and C. Naramoto, *ibid.*, **62**, 163 (1971).
22. J. B. Vos and Th. de Roij, *Toxicol. Appl. Pharm.*, **21**, 549 (1972).
23. C. Hirayama, N. Nakamura and M. Yoshinari, *Fukuoka Acta Med.*, **63**, 405 (1972).
24. T. Yamamoto, C. Hirayama and T. Irisa, *ibid.*, **62**, 85 (1971).
25. M. R. Juchau, T. E. Gram and J. R. Fouts, *Gastroenterology*, **51**, 213 (1966).
26. R. W. Risebrough, P. Rieche, D. B. Peakall, S. G. Herman, and M. N. Kirven, *Nature*, **220**, 1098 (1968).
27. J. G. Vos, J. J. T. W. A. Strik, C. W. M. van Holsteyn and J. H. Pennings, *Toxicol. Appl. Pharm.*, **20**, 232 (1971).
28. C. Hirayama, M. Okumura, J. Nagai and Y. Masuda, *Clin. Chim. Acta*, **55**, 97 (1974).
19. H. Uzawa, Y. Ito, A. Notomi and S. Katsuki, *Fukuoka Acta Med.*, **60**, 449 (1969).
30. M. Okumura, Y. Masuda and S. Nakamura, *ibid.*, **65**, 84 (1974).
31. H. Uzawa, Y. Ito, A. Notomi, S. Hori, Y. Ikeura and S. Katsuki, *ibid.*, **62**, 66 (1971).
32. M. Kikuchi and Y. Masuda, *Jap. J. Clin. Path.*, **21**, 422 (1973).
33. I. Funatsu, F. Yamashita, T. Yoshikane, T. Funatsu, Y. Ito, S. Tsugawa, M. Hayashi, T. Kato, M. Yakushiji, G. Okamoto, A. Arima, N. Adachi, K. Takahashi, M. Miyahara, Y. Tashiro, M. Shimomura, S. Yamasaki, T. Arima, T. Kuno, H. Ide and I. Ide, *Fukuoka Acta Med.*, **62**, 139 (1971).
34. T. Yoshimura, *ibid.*, **65**, 74 (1974).
35. A. Yamaguchi, T. Yoshimura and M. Kuratsune, *ibid.*, **62**, 117 (1971).
36. A. H. Conney, R. Welch, R. Kuntzman, R. Chang, M. Jacobson, A. D. Munro-Faure, A. W. Peck, A. Bye, A. Polans, P. J. Poppers, M. Finster and J. A. Wolff, *Ann. N.Y. Acad. Sci.*, **179**, 155 (1971).
37. Y. Masuda, R. Kagawa and M. Kuratune, *Bull. Environ. Contam. Toxicol.*, **11**, 213 (1974).
38. Y. Masuda, R. Kagawa, K. Shimamura, M. Takada and M. Kuratsune, *Fukuoka Acta Med.*, **65**, 25 (1974).
39. M. Takamatsu, Y. Inoue and S. Abe, *ibid.*, **65**, 28 (1974).
40. M. R. Zavon, *J.A.M.A.*, **190**, 595 (1964).
41. K. Tanaka and Y. Araki, *Fukuoka Acta Med.*, **65**, 53 (1974).
42. C. Hirayama, *Minophagen Med. Rev.*, **15**, 195 (1970).
43. W. E. Dale, T. B. Gaines and W. J. Hayes, *Toxicol. Appl. Pharm.*, **4**, 89 (1962).
44. M. Imamura, *Fukuoka Acta Med.*, **63**, 412 (1972).

7

The Dermal Symptomatology
of Yusho

Harukuni URABE
and Hiromu KODA

7.1 INTRODUCTION

As described in Chapter 1, a 3-yr-old girl visited the Out-patient Clinic, Dept. of Dermatology, Faculty of Medicine, Kyushu University on June 7, 1968, with an acneform eruption as her chief complaint. In early August, her parents and elder sister also came to the clinic complaining of similar symptoms, and were provided with appropriate medical attention. Furthermore, three other families visited the same clinic, so giving rise to the suspicion that the disease might be breaking out on a large scale.

Since the dermal findings resembled those of chloracne, the Dept. of Dermatology, Kyushu University suspected that organic chlorine or some agricultural chemical could have been ingested by the patients, perhaps mixed with rice oil, which was a common element of the diet of affected families. On October 14, a Study Group for the disease[1] was organized with the Faculty of Medicine, Kyushu University playing a leading role, and an Out-patient Clinic was set up to initiate full medical consultations and treatment. On October 19, details of diagnostic criteria for the disease (termed "Yusho") were published, and on this basis medical examinations were conducted in various parts of Fukuoka Pref. in order to clarify the actual state of affairs surrounding the outbreak.

In this chapter, the dermal findings of Yusho patients are described collectively, centered on the 138 patients visiting the Out-patient Clinic up to January 20, 1969. Furthermore, since medical examinations have subsequently been carried out once a year in a follow-up study, mention will be made of the clinical course taken thereafter.

7.2   Clinical Symptoms

The principal dermal findings consist mainly of those based on follicular keratosis, and present a varied clinical picture including dry skin, marked enlargement and elevation of the follicular orifice, comedo formation and acneform eruptions. Differences were seen according to the amount of PCB-contaminated oil ingested, patient age, and lesion site.

For the purpose of describing the course of onset of the dermal symptoms, mention will first be made of one patient, reported by Goto et al.,[1] who at present continues to receive medical treatment:

CASE Y. M., aged 33, female (see Fig. 7.1).
PAST HISTORY: not remarkable.
FAMILY HISTORY : similar symptoms have been observed in all other family members (husband, two children). The patient purchased Kanemi rice oil jointly with several other nearby families at the end of February; similar symptoms were observed in all members of these families.

COURSE OF PRESENT SICKNESS : suffering from edematous swelling of the upper eyelid, hypersecretion of the meibomian glands and amblyopia, which developed from around the end of March, 1968, the patient received treatment from an oculist in her neighborhood without satisfactory results. Nail pigmentation became evident from the end of April, followed by pigmentation of and marked enlargement and elevation of the follicular orifices in the axillae and groin, from the beginning of May. Comedolike eruptions soon developed on the face, and on July 7, the patient had a stillbirth in the 8th month of pregnancy. The skin of the stillborn child is said to have been abnormally blackish. With the symptoms becoming increasingly exacerbated, the patient visited the Out-patient Clinic, Dept. of Dermatology, Kyushu University, for medical consultation and treatment on August 9.

PRESENT SYMPTOMS : the face has become swollen all over in an edematous manner, and a number of black comedones are present in the che-

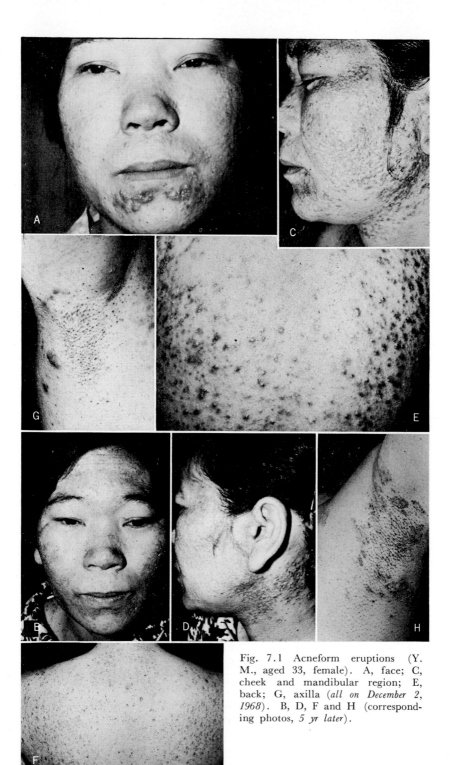

Fig. 7.1 Acneform eruptions (Y. M., aged 33, female). A, face; C, cheek and mandibular region; E, back; G, axilla (*all on December 2, 1968*). B, D, F and H (corresponding photos, *5 yr later*).

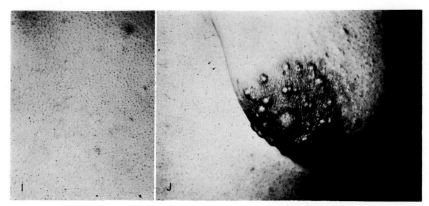

Fig. 7.1—*Continued* I, enlargement and elevation of follicular orifices on the breast; J, hypertrophy of Montgomery's glands.

ek, upper lip and chin.   Those in the chin are accompanied by inflammatory symptoms to form red plaques.   The tip of the nose shows a dirty brownish-blue color, which is specific.   Edemas are particularly prominent in the upper eyelids, while the palpebral conjunctiva and bulbar conjunctiva have become cloudy, and material of cream-cheese-like consistency is being pushed out from the meibomian glands.   A marked enlargement and elevation of the follicular orifices in the neck, back, breast, axillae and groin is apparent, and brown pigmentation is visible in the finger- and toe-nails.

CLINICAL COURSE :   the patient received general treatment for acne while the cause of her disease remained unknown.   However, the symptoms still became exacerbated, and by November prominent acneform eruptions were present even in the ear lobe and external ear canal, and particularly in the retroauricular region.   Moreover, depressed scars were formed in the cheeks after discharge of the contents of the comedones, resulting in a marked deterioration of looks.   Cyst formation occurred in the back, breast and genitocrural region, and this, complicated with secondary infections, yielded a large number of suppurative foci.

The following sections describe each of the major dermal symptoms of Yusho in detail.

7.2.1 Acneform eruptions[1,2]

Acneform eruptions were observed in 113 (81.7%) of the 138 patients examined.

Based on the comedones, the anceform eruption consists primarily of black comedones or cysts about the size of a pea.   Often complicated by secondary infections, this gives rise to inflammatory symptoms, showing a general aspect similar to acne vulgaris.   In serious cases, there is also a tendency for infectious atheroma-like abscesses to form.

Acneform eruptions develop primarily in areas where there is active secretion of sebum.   In the face, the sites of development center around the cheeks or malar region as in acne vulgaris.   However, they may also be located in the glabella, chin, mandibular region, ear lobe, retroauricular region and external ear canal.   In particular, as in the example a-bove, the ear lobe, retroauricular region and external ear canal show a predisposition to acneform eruptions, next only to the face, and so serve as a reliable diagnostic feature of the disease.   In addition, acneform erup-tions occur over the area from the axillae, scapular region, and umbilicus to the pelvic region.   They also develop in the external genitalia, such as the penis (Fig. 7.2) and scrotum, and buttocks, where they often form large cysts. These, coupled with secondary infection, may yield a pilonidal sinus.

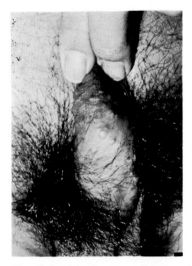

Fig. 7.2 Acneform eruptions on the penis (K. K., aged 18, male).

The above symptoms are conspicuous in patients from puberty to adolescence and adulthood, but rather mild in elderly persons.   In chil-dren, the eruptions are highly characteristic, i.e. uniform black comedones tend to develop in the cheek in a circumscribed manner (see Fig. 7.3A).

There is no tendency for the eruptions to heal spontaneously.   In the case of comedones, healing occurs with discharge of the keratotic plug or keratinous material, leaving concave scars.   In the case of eruptions with accompanying cyst formation, there is no known method of treatment but to extirpate them surgically.

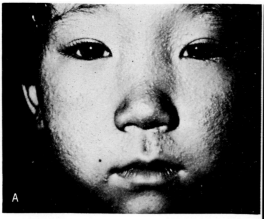
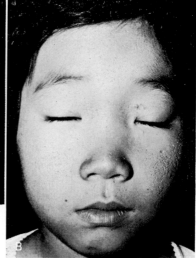

Fig. 7.3 A, Acneform eruptions on the face of a child (K. M., aged 7, female); B, 5 yr later.

Histopathological findings: from the follicular orifice to the entire hair follicle a cyst is formed which is filled with layered keratinous material (see Fig. 7.4).   The sebaceous glands atrophy under pressure and are mostly invisible.   With the walls broken, some have the appearance of a foreign-body granuloma (Fig. 7.5).   The epidermis adjacent to the orifice shows an increase in melanin in the basal layer, and hyperkeratosis; however, no acanthosis is observed.

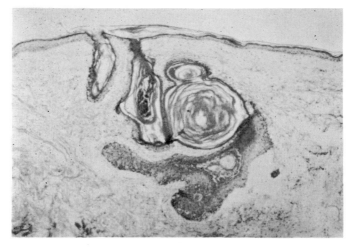

Fig. 7.4   Intrafollicular hyperkeratosis.

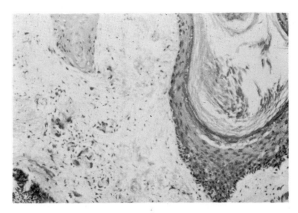

Fig. 7.5 Foreign-body giant cell reaction.

7.2.2 Marked enlargement and elevation of the follicular orifice[1,2]

This condition (Fig. 7.1 I) is characteristic of the disease. It occurs as a normal-colored or blackish fine spotty papule coincident with the follicular orifice. It was seen in 94 (70.1%) of 134 patients.

By region, the condition is most common in the axillae, groin, and glenoid regions, such as at the elbow and knee flexures, followed by the trunk (mainly in the interscapular region and front of the chest) and the extensor aspect of the thigh and outer aspect of the forearm.

The papules are often black in the glenoid regions, but tend to have a normal color on the extensor aspect of the thigh and outer aspect of the forearm. In the latter case, the condition resembles pilar keratosis.

Such eruptions are found in all age brackets. While they are less conspicuous in adolecents and adults, due to the dominance of acneform eruptions, they do represent a prominent symptom in children.

Histopathological findings: hyperkeratosis and formation of keratotic plugs (see Fig. 7.6) are visible at the follicular orifice. The infundibulum

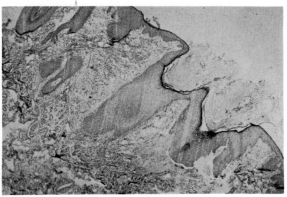

Fig. 7.6 Formation of keratotic plugs.

is tulip-shaped and has a stratified horny plate and spiral vellus hair inside.

7.2.3 Dry skin[1,2]

This symptom is observable mainly in children.   Typically, the skin is dry, showing branny scaling.   If accompanied by nodules coincident with the follicular orifices, it resembles atopic dry skin.   A tendency for the skin to become dry is also observed in neonates, and may occur to a variable extent in elderly persons over about 50.

Histopathlogical findings: there is hyperkeratosis of the epidermis interspersed with keratotic plugs.   It may be concluded that the tendency towards skin dryness is also attributable to keratotic abnormalities and that both the marked enlargement and elevation of the follicular orifice and acneform eruptions, while ascribable to the same general mechanism, merely represent different clinical states based on a difference in the extent of participation of mature sebaceous glands.

7.2.4 Hyperkeratotic plagues in the palms and soles[1,2] (Fig. 7.7)

In serious cases, there may be localized clavus-like hyperkeratosis in regions vulnerable to outside stimuli, such as the thenar eminence of the hands, heels, heel-bone and condyles.   This condition resembles that found in the case of arsenic poisoning.

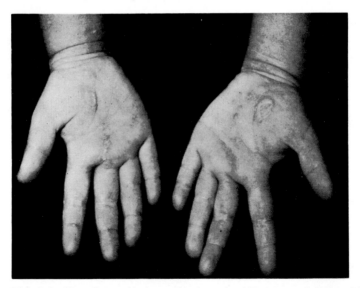

Fig. 7.7   Hyperkeratotic plaques in the palms (K. M., aged 3, female).

7.2.5 Deformation of the nails (Fig. 7.8)

Deformation of the nails has been observed in many cases.   In particular, ingrowing toe-nails may be formed from spicules of the nail of the great toe at the edge of the nail plate.

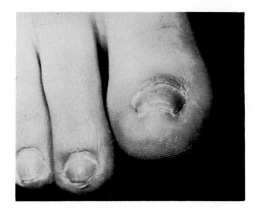

Fig. 7.8 Nail deformation (M. K., aged 24, female).

7.2.6 Changes in hair condition[1]

The hair tends to lose luster, and some patients have complained of an increase in falling hair at the onset of illness.   However, there are no disease-specific features similar to the condition observed with acute toxicosis or infectious and febrile diseases.   There is no record of alopecia totalis or alopecia areata.   Some patients under 10 have shown blackening of the vellus hair on the extensor aspect of the leg.

7.2.7 Pigmentation[1-4] (Figs. 7.9–7.12, see color frontispiece)

Blackish-brown pigmentation of the corneal limbs, conjunctiva, gingivae, lips, oral mucosa and nails is a very specific finding of the disease. Particularly characteristic is the nail pigmentation (Fig. 7.9), which was observed in the toe-nails of 99 (71.7%) and finger-nails of 90 (65.3%) of the 138 patients. In many cases the nails present a general blackish-brown color, although the pigmentation is sometimes in the form of a brown longitudinal streak.   Such nail pigmentation is regarded as an important clinical finding as an initial symptom of the disease, since it is also observed in patients showing no other type of eruption.

In the eyes, pigmentation occurs in both the palpebral conjunctiva and bulbar conjunctiva (Fig. 7.10). Generally, it is prominent in the lower eyelid conjunctiva and lower formix conjunctiva.   In the upper eyelid conjunctiva, it is often present in the nasal half of the palpebral conjunctiva of the upper lids.   Further, it also occurs in the limbal conjunctiva, bulbar conjunctiva and caruncula lacrimalis.

Gingival pigmentation is known at a rate of about 10% even among healthy persons; however, since is was found in almost 70% of the Yusho patients, it was made a finding specific of the disease.   The pigmentation (Fig. 7.11), unlike the wavy pigmentation of the marginal gingivae observed in bismuth poisoning, often presents itself in a belt-like form on the attached gingivae.   Also, there may be an oval or broad-based triangular area of pigmentation in the attached gingivae in the incisal regions.   The pigmentation pattern is thus rather complex and shows a morphological variability.

The lip pigmentation (Fig. 7.12) does not show any consistent form, there being pigmented areas of various shapes, both spotted and diffuse, in the vermilion.   Also, there were several cases where a linear or branch-shaped pigmentation coincident with the occlusal surface of the molars was observed in the oral mucosa.

The above-mentioned types of pigmentation occur irrespective of age and are apparently not related to the severity of other symptoms.   Also, while it is somewhat questionable to define it as pigmentation, the so-called areas of greatest sebaceous activity (particularly centering around the nose) appear as if they have been soiled with dust, and even present a dirty black tone in cases of medium degree or above, which is specific.   This color cannot be removed by washing, and remains almost unchanged even after improvement of the acneform eruptions.   In fact, it can still be seen now after the lapse of several years.

7.2.8 Phymata in the articular regions[1] (Fig. 7.13)

Subcutaneous phymata occur in the elbow, knee and ankle joints.

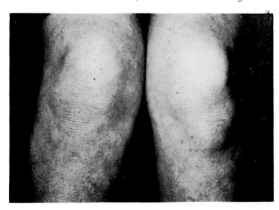

Fig. 7.13 Swelling of the left knee-joint area (T. K., aged 49, female).

They are free of spontaneous pain and give no burning sensation.   Being hemispherical in shape and the size of a pigeon egg to goose egg, they show undulations and retain a clear yellow fluid, giving rise to the possibility that they might be related to the bursa mucosa.   The incidence of such phymata is somewhat higher in middle-aged and older persons.

7.2.9 Dyshidrosis[1,2]

Dyshidrosis can probably be regarded as one of the important features of this disease.   Generally, it presents itself prominently in the form of hyperidrosis of the palms and soles as well as of the glenoidal regions. Beads of sweat are observed in the axillary regions, but there is only a feeling of wet in the palms and soles.

7.2.10. Ocular signs[3]

Complaints about the eyes are the most numerous among the symptoms observed at the onset of illness.   In particular, typical cases complain of such hypersecretion of the meibomian glands that they cannot open their eyes on rising in the morning due to the lid margins being glued down. Also, many individuals are conscious of a swelling of the upper eyelid, amblyopia, diplopia, ophthalmalgia, conjunctival injection, a burning sensation, etc.
Objectively speaking, hypersecretion of the meibomian glands and pigmentation of the conjunctiva are the characteristic findings.   In typical cases, the meibomian glands swell up like a cyst, although the yellow infarct-like contents can be seen through.   As a result, the openings of the glands are raised, the lid margin becomes irregular (Fig. 7.14), and secre-

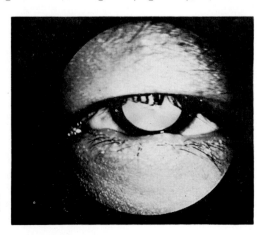

Fig. 7.14 Swelling of the upper eyelid, and irregular lid margin.

tion of material of cream-cheese-like consistency coincident with the meibomian gland openings causes adhesion of the lid margins. Only slight pressure applied to the eyelid results in the white, relatively hard cheese-like contents being pushed out. The cornea is often covered with an oily film due to hypersecretion of the meibomian glands, and many cases show swelling of the eyelids, particularly the upper eyelid (Fig. 7.14).

Hyperemia and opacity are commonly seen in the palpebral conjunctiva, but neither the formation of follicles nor proliferation of the papilla has been observed. The former condition may be an inflammation secondary to the hypersecretion of the meibomian glands, i.e. meibomian conjunctivitis. No organism has been detected from the palpebral sebum.

Such hypersecretion of the meibomian glands shows no difference by sex or age. Further, most cases of the visual disturbance commonly complained of are ascribable to an oily precorneal film and ametropia, which result from hypersecretion of the meibomian glands, there being no actual abnormalities of the cornea, fundus occuli or field or view.

Pigmentation of the conjunctiva, the second main characteristic, has been described earlier (sect. 7.2.7, Fig. 7.10).

7.2.11 Yusho symptoms in neonates[2,5,6]

Mention will next be made of dermal findings made on neonates born to mothers who ingested Kanemi rice oil during pregnancy or who became pregnant after having been diagnosed as Yusho patients.

The skin of such new-born babies typically shows a gray to dark brownish pigmentation over the entire body surface except the palms and soles, and the skin is dry from lamellar scaling. The pigmentation is particularly prominent in the terminal regions of the hands and feet and is found also in the gingivae and corneal limbs. The eylids become swollen edematously (Fig. 7.15, see color frontispiece) and there is hypersecretion of the meibomian glands.

As regards the teeth, irregular hypertrophy of the gingivae and early eruption of infantile incisors occur (see Fig. 7.16). However, with new-born babies, the teeth come off in about one month and the pigmentation also disappears within 3-4 months.

Histopathological findings: this paragraph briefly covers dermal findings made on autopsy of still-born babies with Yusho. In the case of the skin, epidermal hyperkeratosis is apparent over the entire body surface. The epidermis (apart from the horny layers) is atrophic, a condition which is particularly conspicuous in the prickle cell layers. Melanin deposition is prominent in all basal cell layers except on the palms and soles. The openings of the hair follicles are slightly enlarged, and the follicle itself is filled with stratified keratinous material. It is often expanded into a cyst-like structure, a condition which is most prominent on the head. Hyper-

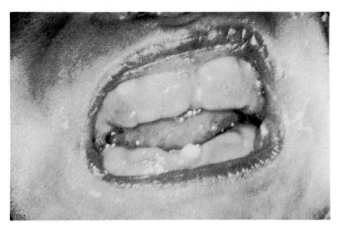

Fig. 7.16 Irregular gingival hypertrophy and early eruption of an in-fantile incisor.

trophy of the sebaceous glands is rarely seen, and there are no marked changes of the sweat glands.

New-born babies with Yusho are thus fundamentally similar to gen-eral Yusho patients as regards their dermal changes.   Specific features are that melanin deposition occurs uniformly in the skin over the entire body surface, that hyperkeratosis is seen widely in the epidermis, and that hy-perkeratosis of the hair follicles occurs prominently in the head.

7.2.12 Grading according to degree of severity of symptoms[1]

The above account essentially describes the individual dermal symp-toms of Yusho; however, this is not to say that all symptoms are found in any one patient. Naturally, there is considerable variability in the severity of dermal lesions observed, ranging from cases showing only a single symp-tom to those presenting all major symptoms.

Classification of the degree of severity by means of grades is consider-ed useful for grasping the general condition of any particular patient, al-though it should be emphasized that the system described here refers not to Yusho as a whole but simply to the severity of dermal lesions.   Clearly, the degree of severity of Yusho itself should be assessed from an overall judgment of the patient's general condition, including findings related to internal medicine.

The dermal symptoms of Yusho can be classified broadly into four phases, in increasing order of severity, as follows:

GRADE I :   although the meibomian glands become swollen and cheese-like palpebral sebum is discharged on pressure, there are no visible comedones or acneform eruptions.   Pigmentation of the nails is common, while hyperidrosis, pigmentation of the gingivae, and a tendency towards skin dryness are taken as indicative findings.

GRADE II :   in addition to the symptoms of GRADE I, there is clear comedo formation, while follicular keratosis in the articular regions and extensor aspect of the limbs is taken as an indicative finding.

GRADE III :   in addition to the symptoms of GRADES I and II, there are clear acneform eruptions, particularly cysts in the genitocrural regions. Swelling of the eyelids and formation of phymata in the articular regions are taken as indicative findings.

GRADE IV :   in addition to the symptoms of GRADES I–III, there is marked enlargement of the follicular orifices all over the body surface, and the acneform eruptions tend to be distributed widely from the face to the trunk.   High-degree secondary infections are taken as an indicative finding.

When the 138 patients under discussion were classified in this way, GRADE II was the most prevalent (43%), followed by GRADE III (20%) and then GRADES I and IV (both 18%). When viewed with respect to age, and omitting the details, the dermal symptoms of Yusho were apparently relatively light in childhood, of a high degree in many cases from puberty up to the late 30's, but tended to become light again in persons in their 40's or over.

7.3   TREATMENT

Since Kanechlor had been ingested orally in the present Yusho patients, general therapy should naturally constitute the major form of treatment.   However, this aspect is discussed elsewhere in sections on internal medicine (Chapter 6), so that the descriptions here will be confined simply to the treatment of dermal symptoms.

In the case of acneform eruptions, the treatment method used for general acne was employed on a trial basis.   Glutathione, vitamin B_2, vitamin B_6, vitamin E, vitamin A, estrogen, protein anabolic steroids, and linolic acid preparations were given as internal drugs, but they produced few effects worthy of mention.   On the other hand, hexachlorophene and other disinfectants or ointments containing antibiotics (principally external drugs containing keratolytic ingredients) were employed as local therapeutic agents in an attempt to prevent secondary infections.

However, the sulfur drugs had virtually no efficacy.   Vitamin A acid,

althoush strongly irritative, proved effective for removing black comedones, but it was not at all useful for treating comedones which had assumed a cyst-like condition to any degree.   The disinfectants and ointments containing antibiotics employed for preventing infections also proved ineffective, and in the end a general administration of antibiotics was
found necessary.

Those lesions which once became purulent required a surgical procedure, such as drainage by incision or extirpation of the cyst.   Particularly,
those cysts which still remain now after the lapse of several years are mostly
of an atheroma-like form, and a surgical procedure is the only treatment
course available for them.

At present, another important therapeutic problem is posed by the
fact that the acneform eruptions and comedones leave depressed scars after healing, and those in the face present an ugly aspect resembling that
remaining after the cure of smallpox.   While dermal abrasion applied to

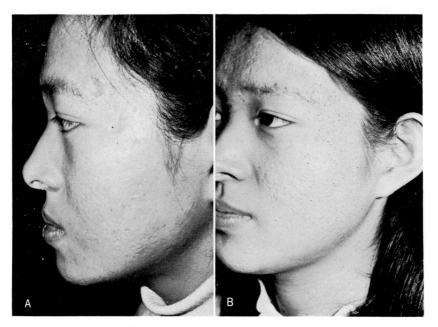

Fig. 7.17 Effect of dermal abrasion (M. K., aged 19, female). A,
before treatment; B, after 1 yr treatment.

the face may produce a fairly satisfactory result (Fig. 7.17), new methods
of treatment are required.

Symptomatic therapies, e.g. vitamin A or E ointment for dry skin and
lichen pilaris-like eruptions, and salicylate plaster for clavus-like hyperkeratosis, have brought about primary symptomatic improvements.   How-

ever, steroid ointments have proved virtually ineffective.

For the various types of pigmentation, vitamin C, glutathione and placenta extracts were employed. However, they showed no corrective action.

To relieve ocular effects, the application of aqueous eye lotions containing antibiotics or sulfa drugs, and drainage of the sebaceous contents of the miebomian glands by light massage, were performed. However, they were successful only in producing symptomatic improvements.

As special therapy, balneotherapy was performed on certain of the patients, at the Institute of Balneotherapeutics, Kyushu University.[7]   Only serious cases were selected for this treatment, and a simple spring was used. The patients were instructed to take a bath 1–2 to 3–4 times a day, lasting about 10 min per time. For individuals whose general condition was good, an acidic hydrogen sulfide spring was mainly used, and they were instructed to take 2–5 baths a day.   Examinations of patients undergoing such therapy for 30 days or more showed that the contents of their abscesses and comedones were more easily dischargeable after treatment, and the condition of the eruptions themselves improved.   It is thus hopefully surmised that some pharmacological effect of the sulfur spring, in addition to the beneficial effect of taking a hot bath, produced these favorable results.

7.4   CLINICAL COURSE

As mentioned above, in addition to organizing the Study Group for Yusho and setting up an Out-patient Clinic in the general out-patient department of Kyushu University Hospital for medical consultations and treatment of the disease, the Faculty of Medicine, Kyushu University has since conducted simultaneous medical examinations of Yusho patients once a year to determine fluctuations in their symptoms. About five years have now passed since this work was initiated.

In the first two or three years, the dermal symptoms showed a marked improvement, to such an extent as to leave only slight pigmentation of the gingival and other areas, in the group represented by light cases of GRADES I and II.   However, in the group represented by serious cases of GRADE III, and particularly of GRADE IV, a strong tendency for cyst formation and for repeated infection of these cysts was observed.   Thus, it can be said that only a limited amount of real progress has been made on the whole (cf. Fig. 7.1A,C,E,G with B,D,F,H).[8]

In serious cases, the dirty black color tone centering around the tip of the nose, and nail pigmentation are still prominent.   In the case of ocular symptoms, the palpebral edemas have disappeared, but many patients still continue to complain of hypersecretion of the meibomian glands, light as it is.

7.5 Mechanisms

Following the report of Herxheimer in 1899,[9] so-called "chloracne" was discussed chiefly in relation to the patients' occupation. On the other hand, since Wauer reported the occurrence of acneform eruptions due to chlornaphthalene in a gas-mask manufacturing factory in World War I (1918), and termed the disease "Pernakrankheit" after the causative a-gent, organic chlorine has come to attract attention as a causative substance.

Although known cases of chloracne hitherto reported are generally the result of contact with poisonous substances from the outside, the present chloracne-like disease (i.e. Yusho) is characterized by the fact that it was caused by the oral ingestion of a poisonous substance. The only similar case is that reported by Herzberg in 1947, [10] who described a family that had suffered from fried potato-poisoning due to the ingestion of chlor-paraffin.

As regards the mechanism by which externally-induced chloracne occurs, an experiment has been conducted by Shelley et al.[11] By application of an ointment containing naphthalene chloride (50%) to human skin, they successfully produced acneform eruptions in the face, back axillae, scalp, scrotum, auricle, and under limbs. They also stated that it was the direct action of the drug on the hair follicle and epidermis which caused follicular keratosis, and that this further led to the formation of keratic cysts. Changes in the sebaceous glands seen in the tissue picture were regarded as secondary changes.

As described above in section 2, the principal dermal symptoms of Yusho consist of skin dryness based on follicular keratosis, marked enlargement and elevation of the follicular orifice, comedo formation, and acneform eruptions. These findings are almost identical with what has previously been described as " chloracne". However, for a poisonous substance ingested orally to present the same clinical picture and tissue effects as those of one applied from the outside, it would be necessary for the poisonous substance to be excreted from the body and yet to act directly on the hair follicles. That is to say, the process is explicable if the ingested Kane-chlor was first excreted into the hair follicles through the sebaceous glands and then stimulated the follicular epidermis directly. This theory is apparently substantiated by the fact that Kanechlor has been found in high concentrations in the sebum and subcutaneous fatty tissue of patients with Yusho (i.e. 30–50 ppm (in terms of Kanechlor 400) for the former and 75.5 ppm[1] for the latter) and also by the fact that the strongest symptoms are observed clinically in puberty to adulthood, when the function of the sebaceous glands is most active, and are most noticeable in the seborrheic regions such as the face, axillae, umbilicus and genitocrural regions.

Further, the histological finding that the sebaceous glands are still clearly observed in early foci, but are atrophic in a completely formed cyst, is in agreement with the results of the human experiments conducted by Shelley et al.[11] This could possibly be interpreted as showing that the Kanechlor is first excreted into the hair follicle without destroying the sebaceous gland, that it then stimulates the hair follicle directly to form keratotic cysts, and that these secondarily press on and cause atrophy of the sebaceous gland.

There is of course certain justification for the view that the dermal symptoms of Yusho are a kind of acute toxicoderma resulting from, and ultimately ascribable to, internal processes following the oral ingestion of Kanechlor. However, objectively speaking, it seems better to regard the symptoms as being produced by an external agency, i.e. the external action and stimulatory effect of a substance as it is excreted from the sebaceous glands.

Regarding the possibility that Kanechlor 400 might also induce keratic cyst formation after external application, we have confirmed that it is able to produce keratotic cysts in rabbits by treating the ear lobe with 20% Kanechlor vaseline. However, we have not as yet conducted any similar experiments on humans.

7.6   CONCLUSION

In general, the degree of severity of the dermal findings naturally depends primarily on the amount of Kanechlor ingested. However, the degree of development of the sebaceous glands, or the age factor, and any predisposition to comedo formation are considered as important additional factors. In other words, there may be cases where acneform eruptions are barely developed despite the ingestion of large quantities of Kanechlor. We would also like to emphasize that grading according to the severity of dermal symptoms does not necessarily correspond to the grading according to the overall severity of Yusho.

Although more than 5 yr have passed since the outbreak of Yusho, not one effective form of therapy has yet been found. The only real consolation is that the early probe into the nature of the causative substance prevented the occurrence of more serious cases or a mass outbreak.

In order to relieve the dermal symptoms, various therapies for general acne were tried. However, they all proved to have little or no effect. Only vitamin A acid was useful for treating black comedones. Even a general administration of antibiotics for the prevention of infections or treatment of comedones or cysts had little beneficial effect, probably due to a defective mechanism for preventing infections arising from the systemic influence of Kanechlor.   Indeed, it sometimes led to complicated cases, with infections by *Proteus vulgaris* appearing after long-term drug administration.

Concerning the surgical extirpation of cysts, it is of course practically impossible to remove all of them in cases where a large number has developed.   Thus, in conclusion, it may be said that this disease still leaves many unsolved questions from the viewpoint of dermatology, not least of which is the development of effective plastic treatment for improving facial appearance that is made ugly by the presence of depressed scars.

REFERENCES

1. M. Goto and K. Higuchi, *Fukuoka Acta Med.*, **60**, 409 (1969).
2. K. Nakashima *et al.*, *Nishinihon J. Derm.*, **33**, 360 (1971).
3. H. Ikui *et al.*, *Fukuoka Acta Med.*, **60**, 432 (1969).
4. M. Aono and H. Okada, *ibid.*, **60**, 468 (1969).
5. M. Kikuchi *et al.*, *ibid.*, **60**, 489 (1969).
6. I. Funatsu *et al.*, *ibid.*, **62**, 139 (1971).
7. Y. Nakamizo and T. Saruta, *ibid.*, **62**, 176 (1971).
8. H. Koda *et al.*, *ibid.*, **65**, 81 (1974).
9. K. Herxheimer, *Münch. Med. Wschr.*, **46**, 278 (1899).
10. H. J. Herzberg, *Derm. Wschr.*, **7**, 425 (1947).
11. W. B. Shelley and A. M. Kligman, *A. M. A. Arch. Derm. Syph.*, **75**, 689 (1957).

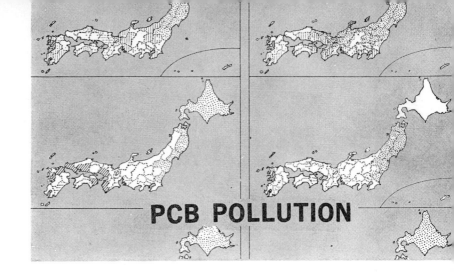

PCB POLLUTION

8

PCB Microanalysis

Hiroya TANABE

In 1971, the Japanese Government established microanalytical methods for PCB's accumulating in foods due to environmental pollution. Tolerance levels for food PCB's were then established on the basis of a large amount of data obtained by these methods. Most kinds of microanalytical techniques for other environmental PCB's were also derived from the methodology for foods. An outline of this methodology is given in the following sections.

8.1 APPARATUS

The basic apparatus is a gas chromatograph equipped with an electron capture detector (ECD).

8.2 REAGENTS

All kinds of organic solvents used are required to pass the following test: first, concentrate 300 ml solvent under suction to 5 ml in all glass apparatus; then, subject 5 µl of the concentrate to ECD-gas chromatography.   The heights of the peaks in the chromatogram due to substances other than the solvent should be less than that corresponding to 2×10^{-11} g γ-BHC.

The distilled water to be used should be so pure that its n-hexane washes show no peaks by ECD-gas chromatography.

The Florisil for column chromatography should be activated at 130°C overnight and cooled in a desiccator.   Before use, it should be checked for complete recovery of 0.1 µg each of endrin and dieldrin, following the procedures described below under section 8.4 and section 8.5.

The silica gel (Wako gel S-1) should be activated at 130°C for more than 3 hr (usually overnight).

The antimony pentachloride used should be free of all substances giving PCB-like peaks on an ECD-gas chromatogram.

8.3   STANDARDS

PBC standards are established and prepared by mixing di-, tri-, tetra-, penta- and hexachlorobiphenyls according to the gas chromatogram of a sample solution.   The decachlorobiphenyl used when determining total PCB's should be >98% pure, mp 305–310°C (uncorrected).

8.4   PREPARATION OF SAMPLE SOLUTION

8.4.1   Extraction

(1)   *Animal tissues*   First, weigh out accurately about 10 g sample, add sufficient anhydrous sodium sulfate to absorbe the tissue water and to soften the tissue, and then grind and homogenize.   Place the ground tissue in a centrifuge tube, shake vigorously with 100 ml n-hexane, and centrifuge at 1500 rpm for 5 min.   Transfer the liquid phase to a beaker, and re-extract the remaining tissue twice with 50 ml (each) of n-hexane. Combine all extracts, and concentrate *in vacuo* to remove solvent as completely as possible.   Weigh the remaining fat, and apply 3 g fat to the clean-up procedure (described below).

(2)   *Butter*   Warm and melt the sample at *ca.* 50°C, and filter it through a dry filter paper.   Apply 3 g filtrate to the clean-up procedure, and report the PCB level on a fat basis.

(3)   *Cheese*   Place 25–100 g chipped cheese in a blender, add 2 g

sodium (or potassium) oxalate and 100 ml methanol, and homogenize for 2–3 min.   For samples known to yield troublesome emulsions at this step, add 1 ml water per 2 g sample.   Transfer the homogenized mixture to a 500 ml centrifuge tube, add 50 ml each of ethyl ether and n-hexane, and shake vigorously.   After centrifugation at 1500 rpm for 5 min, transfer the organic layer into a 1 l separatory funnel containing 500–600 ml water. Extract the remaining water layer with two 50 ml portions of an ethyl ether : n-hexane (1 : 1) mixture.   Wash the combined extracts with two 100 ml portions of water, filter the washed extract through a 6 cm anhydrous sodium sulfate column, concentrate and remove the organic solvents completely, and weigh the remaining fat.   Apply 3 g fat to the clean-up procedure.

(4)   *Milk*   First, determine the fat content by Gerber's method. Then, take 150 g sample in a centrifuge tube, and add 150 ml methanol (or ethanol), 2 g sodium (or potassium) oxalate and 75 ml ethyl ether. After shaking vigorously, add 75 ml n-hexane and shake again.   Centrifuge the mixture at 1500 rpm, transfer the organic layer to a 1 l separatory funnel containing 500–600 ml water, and, after mixing gently, discard the water layer.   Wash the organic layer with two further 100 ml portions of water, and filter the organic layer through a 6 cm anhydrous sodium sulfate column.   Concentrate the filtrate completely, and apply 3 g of the remaining fat to the clean-up procedure.

(5)   *Condensed milk and milk powder*   Dilute the sample with (or dissolve it in) a suitable amount of water to yield a regular milk-type sample, and then proceed as with milk.

8.4.2   Preliminary test for PCB's

This is available as a quick test for PCB's only, excluding organochlorine pesticides.   The procedure is as follows.

(1)   *Alkali decomposition*   Accurately weigh out 3 g fat from the sample (or 10 g of unfatty sample) in a 200 ml flask.   Add 50 ml 1-N sodium (or potassium) hydroxide–ethanol solution, and reflux gently for 1 hr in a boiling water bath.   After saponification, cool the reaction mixture to *ca.* 50°C, mix well with 50 ml n-hexane, and cool to room temperature. Transfer the mixture to a separatory funnel completely with 20 ml of a n-hexane : ethanol (1 : 1) mixture, add 25 ml water, and shake vigorously for several seconds.   After allowing to stand, remove the lower layer into another separatory funnel, and extract with 50 ml n-hexane.   Combine this extract with the first n-hexane extract, and repeat the same extraction procedure 3 times.   Combine all the n-hexane extracts obtained, and

wash with three 100 ml portions of water previously washed with *n*-hexane. Filter the washed extract through an anhydrous sodium sulfate column, and concentrate the filtrate in a Kuderna-Danish evaporator to a definite volume (<5 ml).

(2)   *Clean-up by silica gel chromatography*   Place 2–3 g activated silica gel in a beaker containing about 25 ml *n*-hexane, mix well with a glass rod to remove air, pour the mixture into a 1×30 cm glass tube with an attached cock, and prepare a silica gel column in the regular way.   Then, place about 1 g anhydrous sodium sulfate on the column.   After washing the column with about 20 ml *n*-hexane, apply the PCB extract to it. Allow the extract solution to flow through the column until its surface comes close to the top of the column, and wash the inner wall of the glass tube with several 2 ml portions of *n*-hexane.   Then, allow the solution to flow through the column, and elute with *n*-hexane.   Collect the 50th–100th ml fraction, concentrate the effluent in a Kuderna-Danish evaporator to a definite volume (<5 ml), and subject it to ECD-gas chromatography.

(3)   *ECD-gas chromatography for PCB's*   Proceed according to the analytical technique described below (section 8.5.1).   In the case of analysis for PCB's only (excluding organochlorine pesticides), also determine the total PCB's according to the decachlorobiphenyl method (see section 8.2.5).

8.4.3   Clean-up of organochlorine compounds (PCB's and organo-chlorine pesticides)

(1)   *Partitioning of organochlorine compounds between acetonitrile and n-hexane*   Place 3 g fat sample in a 125 ml separatory funnel with a small amount of *n*-hexane, dilute the solution to 15 ml with *n*-hexane, add 30 ml acetonitrile saturated with *n*-hexane, and shake vigorously for 1 min. Transfer the acetonitrile layer to a 1 l separately funnel containing 700 ml 2% sodium chloride solution and 100 ml *n*-hexane.   Wash the *n*-hexane layer in the 125 ml separatory funnel with three 30 ml portions of aceto-nitrile saturated with *n*-hexane.   Combine all acetonitrile extracts in the 1 l separatory funnel, mix well, and allow to stand until complete separa-tion is achieved.   Transfer the water layer to a second 1 l separatory fun-nel, and extract with 100 ml *n*-hexane by shaking vigorously for 15 sec. Combine the *n*-hexane layers from both 1 l separatory funnels, and wash with three 100 ml portions of 2% sodium chloride solution.   Filter the organic layer through a 6 cm anhydrous sodium sulfate column, concen-trate the filtrate to 10 ml in a 500 ml Kuderna-Danish evaporator, and subject the concentrate to Florisil column chromatography.

(2)   *Florisil column chromatography*   Prepare the column by packing

activated Florisil (with *n*-hexane) 12 cm high in a 2 × 30 cm glass tube with an attached cock, in the regular way.   Apply the material obtained in (1) to the column, wash the container of this *n*-hexane solution with two 5 ml portions of *n*-hexane, and add all washings to the column.   Wash the inner wall of the glass column with a small amount of *n*-hexane.   Allow most of the *n*-hexane solution to flow through the column, then elute organochlorine compounds with 200–300 ml 15% ethyl ether in *n*-hexane, concentrate the effluent to <4 ml, and subject this concentrate to silica gel chromatography.

8.4.4   Separation of PCB's and organochlorine pesticides

(1)   *Rough estimation of the PCB : DDE ratio, and separation of PCB's from other organochlorine pesticides in samples of high DDE content*   Prepare a silica gel column by packing 2 g activated Wako gel S-1 (with *n*-hexane) in a 1 × 30 cm glass tube with an attached cock, in the regular way.   Apply 4 ml or less *n*-hexane solution of the Florisil-treated sample obtained above to the column, and allow the solution to flow through until a small amount remains on the top of the column.   Wash the inner wall of the glass column with several 1 ml portions of *n*-hexane.   Fit a separatory funnel containing about 100 ml *n*-hexane to the glass column, open the cock, and elute the organochlorine compounds in the column.   Discard the first 10–20 ml effluent, and then collect the second fraction, which includes all PCB's and DDE but no other organochlorine pesticides.   (The amount of eluent required for obtaining the second fraction should be checked preliminarily; it is generally 50 ml or less.)   Concentrate the effluent collected in the receiver to a definite amount (>1 ml), and subject it to ECD-gas chromatography as described in section 8.5.   Elute the other organochlorine pesticides on the column with 15% ethyl ether in *n*-hexane solution. (The amount of effluent required should again be checked preliminarily.) Concentrate the effluent, and subject it to ECD-gas chromatography.   It should be noted also that if in the ECD-gas chromatogram of the concentrated PCB+DDE fraction, the peak height of DDE is 5 times that of the highest PCB peak, the DDE content can be considered as great enough to be analyzed without separation from PCB's.

(2)   *Complete separation of PCB's from organochlorine pesticides, including DDE*   If the DDE peak height is less than the prescribed value, the DDE should be completely separated from the mixture, for accurate determination, by large-scale silica gel chromatography.   First, prepare a silica gel column by packing 12 g activated silica gel (with *n*-hexane) in a 2 × 30 cm glass tube with an attached cock, in the regular way, and wash the inner wall of the glass tube with *n*-hexane.   Set the flow rate of *n*-hexane through the column at about 2 drops/sec by regulating the cock, and allow it to

flow down until a small amount of the solvent remains on the column. Apply <4 ml of the Florisil-treated sample solution to the column gently. Allow the solution to flow through the column until a small amount of it remains on the column. Then, wash the inner wall of the glass column with several 1 ml portions of n-hexane, fit a separatory funnel containing about 200 ml n-hexane on the glass column, and open the cock. Allow the n-hexane to flow through the column, if necessary under pressure, collecting the effluent in a flask. (No PCB or organochlorine pesticide is included in this first fraction.) When most of the n-hexane has flowed down, place a further 350 ml h-hexane in the funnel, and allow it to flow through the column until a small amount of the solvent remains on the column. Collect this (PCB) fraction in a receiver, concentrate it to a definite volume (>1 ml), and subject the concentrate to ECD-gas chromatography. Then, elute the organochlorine pesticides in the column with 200 ml 30% ethyl ether in n-hexane, concentrate the effluent, and subject it to ECD-gas chromatography. (Preliminary tests should be carried out to determine the activity of the silica gel, and the amounts of eluents required to elute the PCB's and organochlorine pesticides, respectively.)

8.5 MEASUREMENT OF PCB's

8.5.1 Separation analysis by ECD-gas chromatography

(1) *Preparation and selection of the standard PCB mixture* Prepare a standard mixture by employing suitable amounts of Kanechlors (KC-200~KC-600), so that their gas-chromatographic pattern simulates that of the sample. For this purpose the following ratios have been used:

KC-300 + KC-400 (1 : 1)
KC-300 + KC-500 (1 : 1)
KC-300 + KC-600 (1 : 1)
KC-400 + KC-500 (1 : 1)
KC-400 + KC-600 (1 : 1)
KC-300 + KC-400 + KC-500 (1 : 1 : 1)
KC-300 + KC-400 + KC-500 + KC-600 (1 : 1 : 1 : 1)

(2) *Qualitative analysis* For qualitative analysis, ECD-gas chromatography should be employed with several kinds of stationary phase of different polarity, to confirm that the sample chromatogram and standard chromatogram are very similar to each other regardless of the stationary phase used. Examples of the conditions for gas chromatography are as follows:

Carrier: Gas Chrom Q or Chromosorb G or W (acid-washed and silylated)

Stationary phase: 2% QF-1

2% SE-30 (or 2% OV-1, 2% DC-200)

2% OV-17

2% DEGS+0.5% H₃PO₄

Column length: 1.5–2 m

Column temperature: 180–200°C

Carrier gas: N_2

(3)   *Determination of total PCB's by gas chromatography*   Prepare a calibration curve by summing the peak heights of all peaks other than those which overlap with that of DDE in the ECD-gas chromatogram of the standard PCB mixture.   Measure the sum of the peak heights of all peaks obtained from the sample solution whose respective Rt's are almost equal to those of the standard.   Calculate the PCB content of the sample solution based on the calibration curve.

8.5.2   Determination of total PCB's by the decachlorobiphenyl method[1)]

Seal one end of a 1×25 cm hard glass tube, and place the cleaned-up PCB solution (>1 μg PCB) in it.   After evaporating the solvent, add 0.5 ml antimony pentachloride, and seal the other end of the tube.   Heat the reaction mixture in the sealed tube at 200°C for 1 hr.   After the reaction is complete, cool and carefully open the tube.   Transfer the reacted mixture to a 100 ml separatory funnel using 50 ml benzene, and wash successively with 30 ml 30% hydrochloric acid, two 25 ml portions of 10% tartaric acid, and 25 ml water.   Dry the washed benzene solution with anhydrous sodium sulfate, concentrate in a Kuderna-Danish evaporator, and subject the concentrate to ECD-gas chromatography under the following conditions:

Column: (1)   2% DEGS–0/5% H₃PO₄ (relative Rt against DDE= 6.7)

(2)   SE-30 (7.3), or OV-1 (6.8)

Column temperature: 200°C

The approximate total PCB's (relative error $<25\%$) can be calculated from the following equation:

$$\text{Amount of PCB} \fallingdotseq 2 \times \text{Amount of DCB} \times \frac{\text{MW of biphenyl (154)}}{\text{MW of DCB (455)}}$$

Subsequently, the following new chlorination procedure for PCB's has been developed, in which decachlorobiphenyl is obtained by a reaction which can be carried out under atmospheric conditions.

Reagents: anhydrous aluminum chloride

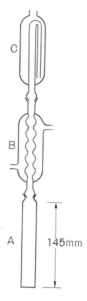

C

B

A   145mm

Fig. 8.1. Apparatus for chlorination of PCB's under atmospheric conditions using anhydrous aluminum chloride, sulfuryl chloride and sulfur
monochloride.

sulfuryl chloride, sulfur monochloried

Apparatus: this is composed of a reaction tube (A), reflux condenser
(B) and gas absorber (C), as shown in Fig. 8.1.   Anhydrous calcium chloride and anhydrous sodium carbo-
nate are packed into the gas absorber.

To the reaction tube containing the concentrated sample solution,
first add a small amount of anhydrous aluminum chloride (vol. equivalent
to a large soybean), quickly crushed in a mortar, and then add 1 ml sulfuryl chloride previously mixed with 1% sulfur monochloride.   Attach
the reflux condenser and gas absorber to the reaction tube, and warm the
tube at 70°C in a water bath for 1 hr.   When the reaction is complete,
remove all excess sulfuryl chloride in the tube under reduced pressure,
re-attach the reflux condenser, and add 5 ml each of water and n-hexane
through the top of the condenser.   Next, remove the condenser, stopper
the reaction tube, and shake it vigorously for 1 min.   Transfer the n-
hexane layer to a test-tube, and after washing with 5 ml 1-N sodium hydroxide, subject the n-hexane solution to ECD-gas chromatography.

In 1973, Kashimoto et al.[2] developed another new, excellent analytical method for PCB's.   They proved by FID-GC and GC-Mass that the
peak areas of pure PCB compounds were almost in inverse proportion to

the numbers of chlorine atoms in the PCB molecules (Table 8.1; Fig. 8.2), and that the components of each one peak on a commercial PCB gas chromatogram were of almost identical chlorine content (Fig. 8.3; Table 8.2).   On this basis, Kashimoto *et al.* experimentally determined a series of correction coefficients for the relation between peak area and amount of PCB's responsible for each peak, i.e. the reciprocal of the relative response of PCB's of certain chlorine content to that of biphenyl (Table 8.1 and 8.2) By application of these coefficients, the PCB amount as percentage (CB%)

TABLE 8.1

FID-Relative Response and Correction Coefficients of PCB Standards

Compound	Formula	Relative response	Correction coefficient[3]
Biphenyl	$C_{12}H_{10}$	100[1]	1.00
2-mono-CB[2] 4-mono-CB	$C_{12}H_9Cl_1$	90.1 88.2	1.13
2,4-di-CB 4,4'-di-CB	$C_{12}H_5Cl_2$	73.3 73.8	1.36
3,4,2'-tri-CB	$C_{12}H_7Cl_3$	65.4	1.53
2,5,2',5'-tetra-CB 2,3,2',4'-tetra-CB 2,4,2',4'-tetra-CB 2,5,3',4'-tetra-CB 3,4,3',4'-tetra-CB	$C_{12}H_6Cl_4$	60.3 56.6 59.5 58.5 57.1	1.71
2,3,4,2',5'-penta-CB	$C_{12}H_5Cl_5$	53.9	1.86
2,4,5,2',4',5'-hexa-CB	$C_{12}H_4Cl_6$	46.5	2.15
2,3,5,6,2',3',5',6'-octa-CB	$C_{12}H_2Cl_8$	40.8	2.45
2,3,4,5,6,2',3',4',5',6'-deca-CB	$C_{12}HCl_{10}$	35.4	2.82

[1] Response to biphenyl=100.   [2] CB=chlorobiphenyl.

[2] Each correction coefficient value for mono-, di- and tetrachlorobiphenyls was calculated as the average of the relative responses of each group.

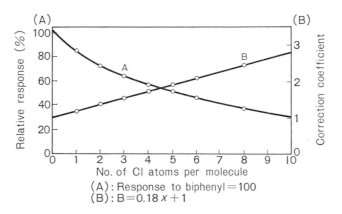

Fig. 8.2.   FID relative response (A) and correction coefficient curves (B).

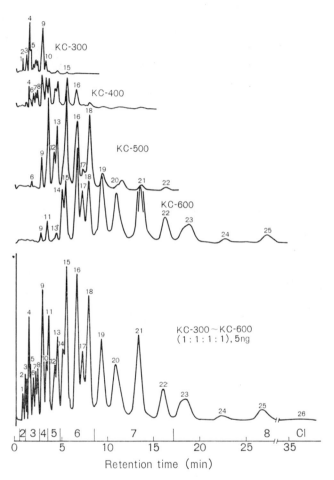

Fig. 8.3. ECD-gas chromatograms of Kanechlors. GC conditions: Varian Aerograph 2100, 2% OV-1 on Chromosorb W (100/200 mesh), 2 mmϕ × 1.5 m glass column; N_2, 45 ml/min; column temp., 170°C; detect. base temp., 250°C; inject. temp., 180°C; sample size, 5 μl ^{3}H·ECD; sensititivity, 2×10^{-9}; chart speed, 10 mm/min.

TABLE 8.2

Determination of the Chlorobiphenyl Contents in Each Peak of
Kanechlor 500 by GC-MS and GC-FID

Peak no.	Peak area[1]	Chlorine atoms	Correction coefficient	Corrected area	CB%[2]
a[3]	402	2	1.36	547	0.10
b					
1	985	2		1340	0.25
c	265	2		360	0.07
2	2915	3		4489	0.88
3	1206	3		1857	0.35
4	3022	3		4654	0.89
5	2086	3		3212	0.61
6	19,227	4	1.72	33,070	6.24
7	5929	4		10,198	1.92
8	2380	4		4094	0.77
9	35,127	5	1.91	67,093	12.66
10					
11	44,366	5	1.91	84,739	15.99
12	22,361	5(6)	1.91	42,710	8.06
13	26,563	5	1.91	50,735	9.57
14					
15	38,867	5(6)	1.91	74,236	14.00
16	28,178	6	2.09	58,892	11.11
17	3554	6	2.09	7428	1.40
18	24,240	6	2.09	50,662	9.56
19	5198	7	2.07	11,799	2.23
20	3957	7		8982	1.69
21	2072	7		4703	0.89
22	1898	7		4308	0.81
	Total			530,108	100.01

[1] Peak area: integrator counts of a hydrogen flame ionization detector.
[2] CB% = content of chlorobiphenyl, calculated as follows:
  peak area × correction coffiecient = corrected area;
  $$\frac{\text{corrected area}}{\sum \text{corrected area}} \times 100 = \text{CB}\%.$$
[3] Peaks a, b and c are significant only in FID-gas chromatograms.

for each peak of the FID-gas chromatogram of a standard PCB preparation obtained under prescribed GC conditions was determined (Tables 8.2 and 8.3). Then, from an ECD-gas chromatogram of the same standard under the same conditions, except for the mode of detection, another series of correction coefficients for the relation between peak height on the ECD-gas chromatogram and amount of PCB responsible for each peak, K in Table 8.4, was calculated (viz) CB%/peak height) (Fig. 8.3; Table 8.4). By application of these coefficients, the PCB content and ratios of PCB's of different chlorine content in samples were then determined under the same GC conditions by measuring the peak heights (Fig. 8.4; Tables 8.5 and 8.6). (The correction coefficients, CB%/peak height, need to be checked

TABLE 8.3
Determined Values of CB% and Chlorine Atoms in Each Peak of
Kanechlors (OV-1 Column)

Peak no.	KC-300		KC-400		KC-500		KC-600		Mixture[2] KC-300~600 (1:1:1:1)	
	Cl no.[1]	CB%[2]	Cl no.	CB%	Cl no.	CB%	Cl no.	CB%	Cl no.	CB%
a	2	1.84	2	0.18	2	0.10				0.53
b	2	0.16							2	0.04
1	2	5.79	2	0.53	2	0.25	2	0.12		1.67
c	3	1.20	3	0.22	3	0.07				0.37
2	3	16.62	3	5.19	3	0.85	3	0.44		5.78
3	3	8.12	3	2.06	3	0.35	3	0.17		2.68
4	3	20.32	3	8.53	3	0.88	3	0.53	3	7.57
5	3	13.40	3	6.51	3	0.61	3	0.38		5.23
6	4	8.62	4	15.82	4	6.24	4	0.84		7.88
7	4	7.53	4	9.46	4	1.92	4	0.41		4.83
8	4	4.53	4	7.53	4	0.77	4	0.38	4	3.30
9	4	7.82	4	19.50	5	12.66	5	2.72		10.68
10	4	2.61	4(5)	6.62			5	0.24		2.37
11	5	0.38	5	4.00	5	15.99	5	2.42		5.70
12	5	0.37	5	4.22	5(6)	8.06				3.16
13	5	0.46	5	4.19	5(6)	9.57	6(5)	2.57	5	4.20
14							6	4.94		1.24
15	5(6)	0.21	5	2.83	5(6)	14.00	6	8.70		6.44
16			5	1.79	6	6.85				2.16
16'					6	4.26	6	11.72		4.00
17			5	0.04	6	1.40	7	5.28	6	1.68
18			5	0.47	6	9.56	6	7.76		4.45
19			6	0.31	7	2.23	7	11.25		3.45
20					7	1.69	7	10.90		3.15
21					7	0.89	7	12.99	7	3.47
22					7	0.81	7	4.28		1.27
23							8	6.16		1.54
24							8	1.16	8	0.29
25							8	2.82		0.71
26							9	0.83	9	0.21
	42.9		47.9		54.6		60.7		Chlorine content(%) [4]	

[1] No. of chlorine atoms per molecule.
[2] Content of chlorobiphenyl   [3] Mean value of KC-300~KC-400.
[4] Calculated values from this table using the formula, $\sum$ CB% × Cl/mol. No. 16
and No. 16' peaks are combined into one peak by the column on chromosorb W.

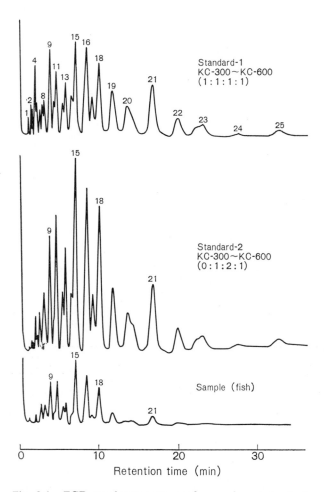

Fig. 8.4. ECD-gas chromatograms of a sample and standards.

Peak no.	Cl no.	Standard PCB (1 ppm·5 μl)			KC-300		
		K[1]	CB%[2]	H[3]	H	M[4]	CB%[5]
1	2	0.186	1.67	6	31	5.8	5.7
2		0.482	5.78	12	39	18.8	(5.8)
3	3	0.206	2.68	13	39	8.0	58.8
4		0.189	7.57	40	99	18.7	(59.7)
5		0.262	5.23	20	55	14.4	
6		0.493	7.88	16	17	8.4	
7		0.284	4.83	17	29	8.2	
8	4	0.194	3.30	17	23	4.5	33.7
9		0.107	10.68	100	99	10.6	(31.1)
10		0.088	2.37	27	30	2.6	
11		0.071	5.70	80	7	0.5	
12		0.096	3.16	33	4	0.4	
13	5	0.076	4.20	55	8	0.6	1.8
14		0.030	1.24	42			(1.4)
15		0.060	6.44	107	5	0.3	
16		0.066	6.16	93	2	0.1	
17	6	0.042	1.68	40	1	0	0.1
18		0.059	4.45	76			
19		0.078	3.45	44			
20	7	0.105	3.15	30			
21		0.075	3.47	46			
22		0.085	1.27	15			
23		0.171	1.54	9			
24	8	0.097	0.29	3			
25		0.142	0.71	5			
26	9	0.210	0.21	1			
Total M		99.1			101.9		
Total PCB (ng)		5.0			5.1		

[1] Correction coefficients for peaks of the ECD chromatogram.
[2] CB% in standard PCB: percentage from the mixture (1 : 1 : 1 : 1) in Table 8.3.
[3] Peak height (mm).
[4] $M = K \times H$ (correction coefficient × peak height).

8.4
for Each of the Kanechlors by GC-ECD

	Sample (1 ppm·5 μl, injected)							
KC-400			KC-500			KC-600		
H	M	CB%	H	M	CB%	H	M	CB%
2	0.4	0.4	1	0.2	0.2			
11	5.3	(0.7)	2	1.0	(0.4)	1	0.5	
9	1.9	20.6	2	0.4	2.9	1	0.2	1.5
45	8.5	(22.5)	4	0.8	(2.8)	3	0.6	(1.5)
18	4.7		2	0.5		1	0.3	
31	15.3		17	8.4		2	1.0	
31	8.8		6	1.7		1	0.3	
40	7.8	58.9	3	0.6	19.6	2	0.4	3.9
196	21.0	(58.9)	70	7.5	(8.9)	24	2.6	(1.6)
62	5.5							
61	4.3		187	13.3		47	3.3	
40	3.8		82	7.9		6	0.6	
54	4.1	16.0	135	10.3	47.8	20	1.5	15.5
		(17.5)			(60.3)	118	3.5	(5.4)
62	3.7		213	12.8		134	8.0	
38	2.5		143	9.4		202	13.3	
2	0.1	3.3	45	1.9	21.9	107	4.5	23.4
12	0.7	(0.3)	153	9.0	(22.1)	132	7.8	(35.7)
3	0.2		36	2.8		140	10.9	
2	0.2	0.3	23	2.4	7.6	207	21.7	45.5
4	0.3		13	1.0	(5.6)	166	12.5	(44.7)
1	0.1		9	0.8		54	4.6	
						42	7.2	
						9	0.9	9.9
						19	2.7	(10.1)
						2	0.4	0.4
	99.2			92.7			109.3	
	5.0			4.7			5.5	

[†5] CB% in sample: calculated value (%) of the PCB of specified number of chlorine atoms. The values in parentheses represent the percentage values of PCB's of specified number of chlorine atom (from Table 8.3).

The total PCB and its composition can be calculated from the table;

for example, total PCB of KC-300 $= 5 \times \dfrac{101.9}{99.1} = 5.1$ ng, and

trichlorobiphenyl in KC-300 $= \dfrac{(18.8+8.0+18.7+14.4)}{101.9} \times 100 = 58.8\%$

Table 8.5

Table 8.5

A Comparison between the Pattern Method and the New Method

Peak no.	Chlorine no.	Standard-1[1]			Standard-2[2]			Sample		
		K	CB%	H	H	M	CB%[3]	H	M	CB%
1	2	0.084	1.67	20	3	0.25	0.2	3	0.25	1.0
2		0.231	5.78	25	9	2.08	(0.3)			
3	3	0.122	2.68	22	7	0.85	7.1			3.9
4		0.085	7.57	89	32	2.72	(7.1)	9	0.77	
5		0.122	5.23	43	13	1.59		2	0.24	
6		0.239	7.88	33	33	7.89		16	3.82	
7		0.107	4.83	45	33	3.53		18	1.93	
8	4	0.063	3.30	52	34	2.14	26.1	17	1.07	41.0
9		0.098	10.68	109	111	10.88	(26.7)	33	3.23	
10		0.049	2.37	48	40	1.96		11	0.54	
11		0.068	5.70	84	139	9.45		36	2.45	
12		0.077	3.16	41	68	5.24		16	1.23	
13	5	0.063	4.20	67	112	7.06	32.5	20	1.26	31.5
14		0.023	1.24	54	75	1.73	(32.3)	11	0.25	
15		0.046	6.44	140	205	9.43		64	2.94	
16		0.048	6.16	129	186	8.93		50	2.40	
17	6	0.029	1.68	57	77	2.23	17.8	10	0.29	16.2
18		0.040	4.45	111	172	6.88	(17.8)	37	1.48	
19		0.051	3.45	67	81	4.13		12	0.61	
20	7	0.072	3.15	44	50	3.60	12.9	4	0.29	5.7
21		0.046	3.47	75	82	3.77	(12.8)	10	0.46	
22		0.049	1.27	26	31	1.52		2	0.10	
23		0.091	1.54	17	27	2.46		2	0.18	
24	8	0.073	0529	4	3	0.22	3.4			0.7
25		0.079	0.71	9	10	0.79	(2.5)			
26	9	0.105	0.21	2						
Total			99.11		1413	101.3		383	25.8	

[1] Standard-1, mixture of KC-300~KC-600 (1 : 1 : 1 : 1).

[2] Standard-2, mixture of KC-400~KC-600 (1 : 2 : 1).

[3] Values in parentheses are calculated values from Table 8.3.

Total PCB (ppm) of sample:

1 (ppm) $\times 5$ (μl) /10 (μl) $\times 25.8/99.1 \times 5$ (ml) /2 (g) $= 0.33$ (ppm) (new method)

1 (ppm) $\times 5$ (μl) /10 (μl) $\times 383/1413 \times 5$ (ml) /2 (g) $= 0.34$ (ppm) (pattern method)

Percentage of tetrachlorobiphenyl in sample:

$(3.82 + 1.93 + 1.07 + 3.23 + 0.54) / 25.8 \times 100 = 41.0\%$.

TABLE 8.6

Total PCB Determinations in Various Laboratories

Sample	Pattern method		New method	Detector	Laboratory
	Mixed ratio of Kanechlors	Total PCB (ppm)	Total PCB (ppm)		
	KC-300, 400, 500, 600				
Crucian	4 : 4 : 4 : 1	1.7	1.7		Osaka Prefectural
Croaker	4 : 2 : 10 : 3	1.8	1.9	^{3}H·ECD	Institute of Public
Sea bream	4 : 2 : 10 : 3	0.23	0.21		Health
Mackerel	0 : 1 : 2 : 1	0.96	0.94		
	KC-200, 500, 600				
STD of					Tokyo Metropolitan
human fat	500 : 1000 : 865	(2.36) †	2.44	^{63}Ni·ECD	Research Laboratory
Human fat	500 : 1000 : 865	1.62	1.42		of Public Health
	KC-300, 400, 500, 600				
Croaker-1	1 : 2 : 2 : 4	0.08	0.08		
Croaker-2	1 : 1 : 2 : 2	0.08	0.10	^{63}Ni·ECD	Nagasaki Prefectural
Bonito, canned	0 : 0 : 1 : 1	0.05	0.03		Laboratory of Public
Greenling	1 : 1 : 2 : 2	0.12	0.10		Health
Saury, canned	1 : 1 : 1 : 1	0.014	0.014		
Goby	1 : 2 : 2 : 2	0.31	0.33		Osaka City Institute
Cuttlefish	3 : 2 : 3 : 0	0.085	0.089	^{63}Ni·ECD	of Hygiene
Meat	1 : 0 : 2 : 0	0.040	0.043		
Turbot-1	1 : 2 : 4 : 1	1.3	1.2		Tokai Regional
Turbot-2	1 : 2 : 4 : 1	0.41	0.42	^{63}Ni·ECD	Fisheries
Turbot-3	1 : 2 : 4 : 1	0.32	0.34		Research Laboratory
Turbot-4	1 : 2 : 4 : 1	0.49	0.56		

† Value in parentheses: theoretical value.

TABLE 8.7
Calculation of Mixed Kanechlors

Mixed ratio of Kanechlors				GC inject. PCB (ng)	Calculation		
KC-300	KC-400	KC-500	KC-600		Total M	PCB (ng)	Error (%)
1	1	1	1	5.0	99†	5.0†	
1	1	1	1	10.0	201	10.2	+ 2
1	1	1	1	15.0	311	15.7	+ 5
1				5.0	101	5.1	+ 2
	1			5.0	99	5.0	0
		1		5.0	94	4.7	− 6
			1	5.0	99	5.0	0
1	1	1		5.0	97	4.9	− 2
1		1		5.0	90	4.5	−10
	1	1	1	5.0	103	5.2	+ 4
1			1	5.0	106	5.4	+ 8
1	1			5.0	108	5.5	+10
1	2	1		5.0	91	4.6	− 8
1		2	1	5.0	105	5.3	+ 6
1	1	2	2	5.0	109	5.5	+10
1		3		5.0	109	5.5	+10
1	3	2	1	5.0	105	5.3	+ 6
	4	1		5.0	107	5.4	+ 8
3	2	1	1	5.0	91	4.6	− 8
6	4	2	1	5.0	109	5.5	+10
	3	1		5.0	92	4.6	− 8
3		1		5.0	93	4.7	− 6
1	3	5		5.0	91	4.6	− 8
3	5	1		5.0	91	4.6	− 8
2	2	1	1	5.0	98	4.9	− 2
5	1	1		5.0	91	4.6	− 8
1	5	2	1	5.0	93	4.7	− 6
5		1		5.0	98	4.9	− 2
10	2	1	1	5.0	94	4.7	− 6
	2	1		5.0	99	5.0	0
	1	2		5.0	101	5.1	+ 2
2	1	1		5.0	107	5.4	+ 8
1	2	3		5.0	101	5.1	+ 2
3	1	1	1	5.0	104	5.3	+ 6
4	1	1	2	5.0	101	5.1	+ 2
		4	1	5.0	107	5.4	+ 8
	1	2	1	5.0	101	5.1	+ 2
	1	1		5.0	108	5.5	+10

† The total PCB (ng) was calculated using the mixture of an equivalent amount of standards (KC-300∼KC-600), 5 ng.

from time to time due to changes in ECD sensitivity.)   As a practical standard, a mixture of equal amounts of Kanechlors (KC-300, KC-400, KC-500 and KC-600) was most suitable.   Moreover, in spite of the fact that due to the small amounts of PCB's of different chlorine contents in each of several peaks, errors of analysis are to some extent inevitable, the real errors of results obtained by this method could generally be maintained within ±10% (Table 8.7).

REFERENCES

1.   T. Mizutani and M. Matsumoto, *J. Food Hyg. Soc. Japan*, **13**, 398 (1972).
2.   M. Ugawa, A. Nakamura and T. Kashimoto, *ibid.*, **14**, 415 (1973); *New Envirnmental Chemistry and Toxicology* (ed. F. Coulston, F. Korte and M. Goto) p. 253, International Academic Printing, 1973.

9

PCB Pollution of the Japanese Environment

Ryo TATSUKAWA

9.1 INTRODUCTION

PCB pollution in the Japanese environment was discovered separately and independently by two groups attached to Kyoto City Hygiene Research Laboratory[1] and Ehime University.[2,3] Following the accidental outbreak of "Yusho" due to oral intake of PCB's in 1968, the successive detection of various amounts of PCB's in the environment severely shocked both the general public and the Administration. To date many surveys and laboratory experiments have been undertaken by the National and Local Governments, universities, various corporations, etc. However, publication of the results obtained has tended to spread over very diverse journals, etc., including non-academic publications and limited-circulation Government papers. The material treated in this chapter is thus limited largely to the results of the author's laboratory in Ehime University.

9.2   PRODUCTION, USE AND PROPERTIES OF PCB'S

According to the Ministry of International Trade and Industry (MITI), the total PCB production in Japan for the period 1954–72 was 58,787 tons, and the total amount used domestically in that period was 54,001 tons (Table 9.1).   The main manufacturer of PCB's in Japan was Kanegafuchi Chemical Co., Ltd.   (96% of the total output for 1954–72), with Mitsubishi-Monsanto Chemical also entering production in 1969. Three-quarters of the total consumption (37,156 tons) was for use in electrical appliances such as transformers and condensers, while 8585 tons were used for heat-transfer agents, 5350 tons for non-carbon copy paper, and 2910 tons for miscellaneous purposes (paints, lubricants, etc.).

Table 9.2 shows the production breakdown on a grade basis for the period 1954–71.   In this table, the PCB's with 3 or less chlorine atoms per molecule (3-Cl or lower PCB's) consist mainly of Kanechlor 300 (KC-300,

TABLE 9.1
Production and Use of PCB's in Japan (Source: MITI, 1972)
(Unit: tons)

| Year | Production in Japan | Imports | Domestic use | | | | | Exports | PCT |
			Electrical appliances	Heat-transfer agents	Non-carbon copy paper	Miscel-laneous	Total		
1953	—	20	—	—	—	—	—	—	—
1954	200	30	200	—	—	—	200	—	—
1955	450	30	430	20	—	—	450	—	10
1956	500	30	430	50	—	20	500	—	20
1957	870	—	760	80	—	30	870	—	20
1958	880	—	740	100	—	40	880	—	30
1959	1260	—	1060	120	—	80	1260	—	30
1960	1640	—	1320	170	—	150	1640	—	50
1961	2220	—	1860	180	—	180	2220	—	70
1962	2190	3	1640	240	10	200	2090	100	60
1963	1810	37	1270	240	30	170	1710	100	100
1964	2670	8	1920	400	100	210	2630	40	180
1965	3000	—	1980	450	170	240	2840	160	90
1966	4410	117	2600	660	300	270	3830	580	150
1967	4480	164	2370	730	390	270	3760	720	260
1968	5130	223	2830	720	780	260	4590	540	260
1969	7730	145	4220	1290	1300	330	7140	590	400
1970	11,110	181	5950	1890	1920	360	10,120	1000	320
1971	6780	170	4560	1160	350	100	6170	730	240
1972	1457	—	1016	85	—	—	1101	758	90
Total	58,787[†1]	1158	37,156	8585	5350	2910	54,001	5318	2380[†2]

[†1] Kanegafuchi Chemical, 56,326 tons; Mitsubishi-Monsanto Chemical, 2461 tons.
[†2] Excluding the production of 350 tons by Mitsubishi-Monsanto Chemical.

TABLE 9.2
Annual Production of PCB's in Japan on a Grade Basis
(Source, MITI, 1972) (Unit: tons)

Year	Production				
	3-Cl or lower PCB's	4-Cl PCB's	5-Cl PCB's	6-Cl or higher PCB's	Total
1954					200
1955					450
1956					500
1957					870
1958					880
1959					1260
1960					1640
1961					2220
1962	860	440	800	90	2190
1963	690	400	640	80	1810
1964	1270	430	880	90	2670
1965	1390	510	980	120	3000
1966	1910	800	1500	200	4410
1967	1860	900	1490	230	4480
1968	2520	850	1560	160	5130
1969	3880	1440	2140	270	7730
1970	5540	1900	3060	610	11,110
1971	2860	1370	2250	300	6,780

Kanegafuchi Chemical) and partly of Aroclor 1242 (AR-1242, Mitsubishi-Monsanto Chemical). Similarly, the 4-Cl, 5-Cl, and 6-Cl or higher PCB's consist of KC-400, KC-500, and KC-600, with smaller amounts of AR-1248, AR-1254, and AR-1260, respectively. The chlorine contents of the series KC-300~KC-600 are the same as those of the series AR-1242, AR-1248, AR-1254 and AR-1260, respectively, but the detailed distributions of PCB isomers in these Kanechlors and Aroclors are different (Table 9.3). The main fields of use of the Kanechlor series are in condensers and non-carbon copy paper (KC-300), heat-transfer agents (KC-400), electrical appliances such as transformers (KC-500), and paints (KC-600).

The total amounts of PCB's shipped to different prefectures in Japan

TABLE 9.3
PCB Compositions (%) by chlorine number[5]

PCB grade	Cl%	No. of Cl atoms/molecule								
		1	2	3	4	5	6	7	8	9
Ar-1242	42	1	16	49	25	9	1			
Ar-1248	48		2	18	40	36	4			
Ar-1254	54				11	49	34	6		
Ar-1260	60					12	38	41	8	1
KC-300	42		6	60	31	1				
KC-400	48		1	22	59	12	6			
KC-500	54			3	21	34	37	8		
KC-600	60			2	5	5	38	39	11	

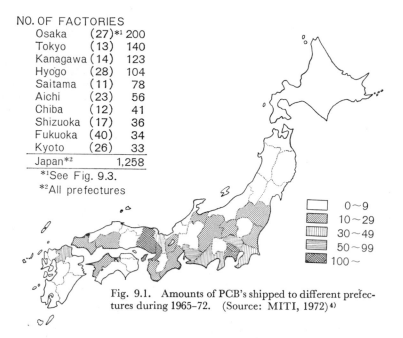

NO. OF FACTORIES

Osaka	(27)*1	200
Tokyo	(13)	140
Kanagawa	(14)	123
Hyogo	(28)	104
Saitama	(11)	78
Aichi	(23)	56
Chiba	(12)	41
Shizuoka	(17)	36
Fukuoka	(40)	34
Kyoto	(26)	33
Japan*2		1,258

*1 See Fig. 9.3.
*2 All prefectures

0~9
10~29
30~49
50~99
100~

Fig. 9.1.   Amounts of PCB's shipped to different prefectures during 1965–72.   (Source: MITI, 1972)[4]

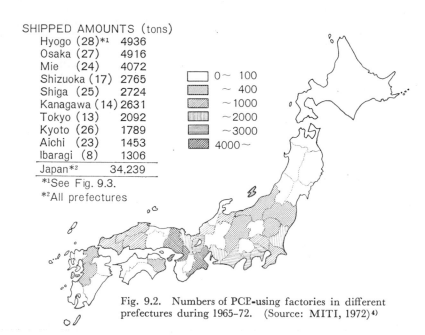

SHIPPED AMOUNTS (tons)

Hyogo	(28)*1	4936
Osaka	(27)	4916
Mie	(24)	4072
Shizuoka	(17)	2765
Shiga	(25)	2724
Kanagawa	(14)	2631
Tokyo	(13)	2092
Kyoto	(26)	1789
Aichi	(23)	1453
Ibaragi	(8)	1306
Japan*2		34,239

*1 See Fig. 9.3.
*2 All prefectures

0~ 100
~ 400
~1000
~2000
~3000
4000~

Fig. 9.2.   Numbers of PCB-using factories in different prefectures during 1965–72.   (Source: MITI, 1972)[4]

Fig. 9.3   Prefectures in Japan.

1	Hokkaido	17	Shizuoka	33	Okayama
2	Aomori	18	Niigata	34	Hiroshima
3	Iwate	19	Toyama	35	Yamaguchi
4	Miyagi	20	Ishikawa	36	Tokushima
5	Akita	21	Fukui	37	Kagawa
6	Yamagata	22	Gifu	38	Ehime
7	Fukushima	23	Aichi	39	Kochi
8	Ibaragi	24	Mie	40	Fukuoka
9	Tochigi	25	Shiga	41	Saga
10	Gunma	26	Kyoto	42	Nagasaki
11	Saitama	27	Osaka	43	Kumamoto
12	Chiba	28	Hyogo	44	Oita
13	Tokyo	29	Nara	45	Miyazaki
14	Kanagawa	30	Wakayama	46	Kagoshima
15	Yamanashi	31	Tottori		
16	Nagano	32	Shimane		

during the period 1965–72 are shown in Fig. 9.1.   The data indicate a very heavy use of PCB's in the Kinki region (prefectures 24–30 in Fig. 9.3). A similar situation is also seen with regard to the number of PCB-using factories in different prefectures (see Fig. 9.2).   The Tokaido belt and Seto Inland Sea regions are the areas of greatest PCB use; in particular, about half of the PCB production was consumed by various factories in the Kinki region.

In the case of pesticides and detergents, the total amount liberated into the natural environment is essentially equal to the total amount produced, and the ban on DDT and BHC production thus means a virtual cessation of their new introduction into the environment after a short time lag.   However, in the case of PCB's, it is estimated that only 20–30% of the total production has been introduced into the environment.   Thus, although PCB production in Japan was stopped in 1972, it is very difficult in practice to terminate completely the utilization of PCB's in all areas.   In addition to the chemical persistence of PCB's, this situation obviously tends to lead to more prolonged pollution.   Isono[4,5] has made an estimate of the total PCB input into the open, natural environment in Japan, as shown in Table 9.4.   According to his conservative estimate, the comparatively small amount of PCB's of about 10,000 tons has been introduced into and accumulated in the environment, although this has clearly had a complex and serious effect on the biota and on society.

Before reviewing the details of PCB pollution in Japan, it is constructive to give a brief comment on the essential physicochemical properties of PCB's, which are listed in Table 9.5.   More highly chlorinated isomers tend to have lower vapor pressures and water solubilities, and are chemically more stable and more persistent in the environment.   Lower chlorine isomers evaporate more easily into the atmosphere and are transferred more easily to an aqueous environment from their source.

TABLE 9.4
Estimated Amount of PCB Input into the Natural Environment and of PCB Recalls[5] (Unit: tons)

| Field of use | Input | | | Recalls |
	Up to the present	In the future	Sum	
Electrical appliances	4000	3100	7100	30,000
Heat-transfer agents	2000	600	2600	6000
Non-carbon copy paper	2700	1700	4400	200~1000
Miscellaneous	1000	1900	2900	0
Total	9700	7300	17,000	~37,000

TABLE 9.5
Some Physicochemical Properties of PCB's (Kanechlors)
(Source: Catalog of Kanegafuchi Chemical Co., Ltd.)

PCB grade	Main components	S. G. (100°C)	Viscosity (75°C, cm^2/sec)	Loss of volatilization (%, 98°C, 5 hr)
KC-200	Dichlorobiphenyls	1.223–1.243	2–3	1.5
KC-300	Trichlorobiphenyls	1.310–1.322	3.5–4.4	0.4
KC-400	Tetrachlorobiphenyls	1.376–1.389	5.4–7.3	0.3
KC-500	Pentachlorobiphenyls	1.460–1.475	12–19	0.2
KC-600	Hexachlorobiphenyls	1.539–1.555	46–87	0.1

PCB grade	Distillation range (°C, 760 mm Hg)	Vapor pressure (35°C, mm)	Solubility in water (room temp., ppm)	Corresponding Aroclor
KC-200	270–360			Ar-1232
KC-300	325–360	0.001	0.147	Ar-1242
KC-400	340–375	0.00037	0.042	Ar-1248
KC-500	365–390	0.00006	0.008	Ar-1254
KC-600	385–420		0.002	Ar-1260

9.3   ENVIRONMENTAL PCB POLLUTION IN JAPAN

9.3.1   Outline of PCB pollution in Japan

The results of various surveys and research programs on PCB pollution in Japan to date may be summarized as follows:

(1)   PCB's have been detected in the whole spectrum of environmental samples (air, water, soil, plants, birds, fish, man etc.), indicating PCB contamination of the entire Japanese environment.

(2)   High levels of PCB contamination exist in the environment of urban and industrial areas.   These are located mainly in the western part of Japan.

(3)   Very high levels of environmental PCB contamination have been found in the vicinity of some transformer/condenser factories and recycling paper-mills.

(4)   Among the various environmental phases, the marine environment and fish of coastal areas are severely contaminated with PCB's.

Fig. 9.4[6] gives a general summary of the main transport routes and contamination levels of PCB's in the Japanese environment.

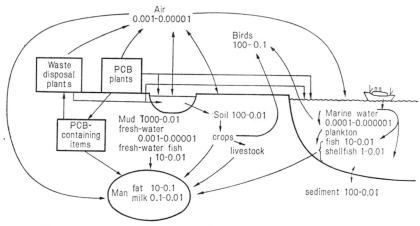

Fig. 9.4.   Fate of PCB's in the environment (unit: $\mu g/g$).[6]   (The numerical values have been modified partly on the basis of recent surveys.)

9.3.2   PCB's in the atmosphere

In 1969, Hara[7] reported measurements of the PCB contamination in the air of an electrical appliances plant (Table 9.6): his report was the

TABLE 9.6

PCB Concentrations in the Air of a Plant Producing Transformers,
Capacitors and Condensers (Before 1957)[7]

Time and place of measurement	PCB (mg/m³)
Oilpan check	4.50
Oilpan, heating	6.75
Taking out of cells	2.13
Soldering test of cells	0.37
Leakage test of cells	2.51

first paper on PCB's in the environment and the only one before 1971. In this old-fashioned fatcory with bad ventilation, very high levels of air contamination were detected and these had caused a number of chloracne patients.

Following resolution of the difficulties of collecting and measuring trace amounts of PCB's from the natural atmosphere by adopting a new collection column packed with glycerin-coated Florisil,[8] air surveys for PCB's were conducted in various parts of Japan (see Table 9.7).   At the

TABLE 9.7

PCB levels in air samples[9]

Station	Date	PCB level (μg/m³)	PCB composition	Remarks
Around PCB-using factories				
Kusatsu City (25) †				
Near a waste-water precipitation pond	June 6-8, 1972	0.60	KC-200	Released grade of PCB's from the factory was mainly KC-300
5 km west of the factory	″	0.03	″	Rainy and cloudy weather
Toyonaka City (27) †				
Above a waste-water conduct from the factory	June 9-11, 1971	12.0	KC-200	Fine weather, KC-300 released
Paper-mill areas				
Fuji City (17) †				
Port Tagonoura	June 7-9, 1972	0.05	KC-300>	KC-300 releared from many vecycling paper-mills
Tanaka Shinden	″	0.05	″	Rainy weather
River Fuji	″	0.12	″	Spreading of PCB-contaminated sediments
Iyomishima City (38) †				
River Chigiri	May 11-13, 1972	0.03	KC-300	KC-300 released from recycling paper-mills
Sangawa	″	0.01	″	
Nakatai	″	0.006	″	

TABLE 9.7—*Continued*

Station	Date	PCB level (μg/m³)	PCB composition	Remarks
Urban areas				
Tokyo (13) †				
Nakano-ku	Aug. 8–10, 1972	0.02	KC-300$\leq$	—
Bunkyo-ku	Aug. 10–11, 1972	0.02	″	
Osaka (27) †				
Toyonaka	June, 1972	0.04	″	—
Osaka City	″	0.02	″	
Matsuyama City (38) †				
Fukuinji	May 24–26, 1972	0.005	″	
Sanbonyanagi	″	0.003	″	
Nishihabu	″	0.003	″	
Tarumi	″	<0.002	″	Roof-top of Ehime University building
Room air				
Warehouses for non-carbon copy paper				
Tokushima Savings Office (27) †	June 17–18, 1972	70	KC-200>	—
Iyosaijo City Office (38) †	July 6–7, 1972	2.0	″	
PCB-using laboratory				
Ehime University (1) (38) †	June 26–28, 1972	0.04	KC-400 and -500	—
Ehime University (2) (38) †	June 28–30, 1972	0.06	KC-300, -400 and -500	

† See Fig. 9.3.

time of 1972, the base-line outdoor PCB levels in the atmosphere were of the order of $0.0n$ μg/m³ in large cities such as Tokyo and Osaka, $0.00n$ μg/m³ in medium-sized cities such as Matsuyama, and below the detection limit (0.001 μg/m³) in rural areas. The air around PCB-using factories such as electrical appliance factories and recycling paper-mills typically showed high PCB levels (up to 12 μg/m³), and the highest level of air PCB's was detected in a stockroom for non-carbon copy paper.

It is interesting that the PCB content of the air is often more abundant in lower chlorine isomers than is that of the volatilization source. For example, in Kusatsu City, waste-water from a condenser factory contained mainly KC-300, whereas the air around the factory contained lower chlorine members than KC-300. Fig. 9.5 demonstrates a similar difference, apparent between the gas chromatograms of sewage bottom-mud and of air above the sewage in the vicinity of an electrical appliance plant in Toyonaka City.

Rapid volatilization of PCB's from the surface of water has been demonstrated by laboratory experiments (see Fig. 9.6).[9] After adding a

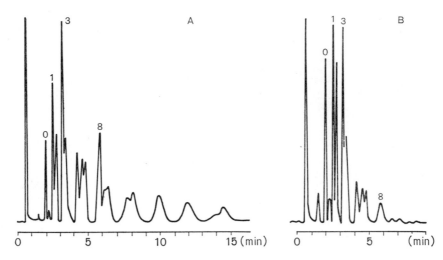

Fig. 9.5.   Gas chromatograms of sewage bottom mud (A) and of air above the sewage (B) (Toyonaka City, Osaka; June, 1972).

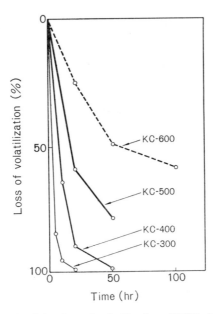

Fig. 9.6.   Loss of volatilization of PCB's from the surface water.[9]

small volume of an acetone solution of PCB's to distilled water in a beaker, the loss of PCB's from the water surface was determined periodically. Within 20 hr, KC-300 (which has a relatively high vapor pressure, see Table 9.5) was almost completely lost from the water, and even in the case of KC-600 (a low vapor pressure grade) more than 50% of the added PCB was lost in 100 hr.

Within each PCB grade, the lower chlorine isomers are more rapidly volatilized, both from liquid and solid surfaces. In the natural environment, oily substances and detergents are known to retard or slow down the volatilization of PCB's from the surface of water. However, the vapor pressure of PCB's is one of the dominant factors determining the fate of PCB's in the environment, and the volatility of PCB's is one of the main causes of their ubiquitous occurrence in the Japanese environment. Fig. 9.7[10] depicts schematically the changes in PCB composition of an air-plant-soil system, based on a survey conducted in the vicinity of a condenser factory in Kusatsu City. As lesser chlorinated isomers volatilize from

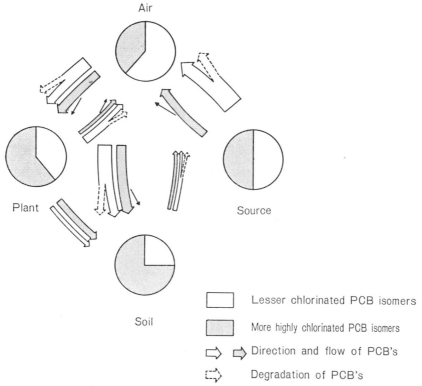

Air

Plant

Source

Soil

▢ Lesser chlorinated PCB isomers

▨ More highly chlorinated PCB isomers

⇨ ⇨ Direction and flow of PCB's

⇢ Degradation of PCB's

Fig. 9.7. Changes of PCB composition in the environment for an air-plant-soil system.[10]

each pollution source, the air accumulates more low chlorinated isomers than the sources.   Vegetation catches the PCB's in the air on its surface, where high vapor pressure isomers (i.e. low chlorinated isomers) tent to evaporate and more highly chlorinated isomers accumulate persistently. In the soil, PCB's introduced as plant debris and as dry and wet precipitation evaporate and are degraded from low chlorinated isomers.   Consequently, there is a process towards gradual concentration of more highly chlorinated isomers in the environment.

9.3.3   PCB's in water

In 1972, the Environment Agency[11] surveyed on a nationwide basis the degree of PCB contamination of water and bottom sediments during the period May through November.   The water quality was tested at 1084 sites, and the deposits at 1445 sites.   As shown in Table 9.8, the PCB concentrations in the water samples were generally low, but the bottom

TABLE 9.8
Results of a Survey on PCB Pollution (Water Quality and Sediments) [11]

PCB content (ppm) Sites	Water quality			Sediments							Grand total
	0.01 or less	0.01 or more	Sub-total	1 or less	1.1 ¿ 10	11 ¿ 50	51 ¿ 99	100 ¿ 499	500 or more	Sub-total	
PCB-using industry	78	13	91	0	19	10	5	11	9	54	145
Sewage treatment plant	125	0	125	—	—	—	—	—	—	—	125
Water near factories	76	4	80	26	24	15	7	1	4	77	157
Public water	785	3	788	1216	73	20	2	3	0	1314	2102
Total	1064	20	1084	1242	116	45	14	15	13	1445	2529

deposits occurring near factories handling PCB's were relatively heavily polluted.

The results of two surveys undertaken in Tokyo Bay and Lake Biwa (Fig. 9.8[12] and 9.9,[13] respectively) will next be discussed briefly.   In the Tokyo Bay area, many PCB-handling factories are located on the northern and western coasts, and they utilize all grades of PCB's.   Thus, both the river and marine sediments in these areas have mixed compositions of PCB's while the eastern part of Tokyo Bay is less polluted with PCB's (Fig. 9.8).   Lake Biwa is the largest freshwater lake in Japan (see Fig. 9.3, no. 25), and several PCB-handling factories are located on its southern shores. Among them, the condenser factory in Kusatsu City, near the Kusatsu River (Fig. 9.9), has been a major source of PCB's polluting the lake.

Following the recent ban on PCB production and expected reduction in PCB utilization, the PCB levels in the terrestrial environment will un-

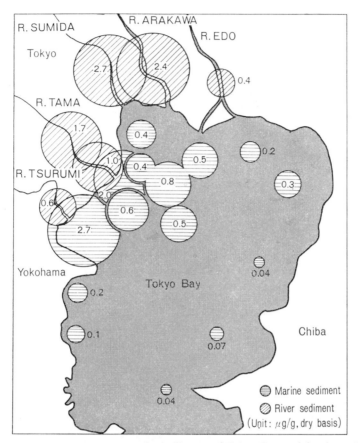

Fig. 9.8.   PCB levels in marine sediments of Tokyo Bay and in the sediments of rivers entering Tokyo Bay (1971).[12]

doubtedly gradually decrease.   However, a certain proportion of the decreased amount will inevitably be transported to the oceans, which represent a final sink for PCB's and numerous other long-life pollutants. Several laboratories are thus currently investigating the distribution and changes of PCB levels (together with those of other organochlorine compounds) in the oceans.

Large-area PCB contamination in the sea water of Suruga Bay (see Fig. 9.3, no. 17) has been detected in a survey conducted by the Shizuoka prefectural authorities (Fig. 9.10).   Many paper mills are located in and around Fuji City, and waste-water highly contaminated with PCB's has been allowed to pour into Suruga Bay from several recycling paper-mills utilizing non-carbon copy paper.

Fig. 9.11 shows the distribution of PCB's in the surface waters in and

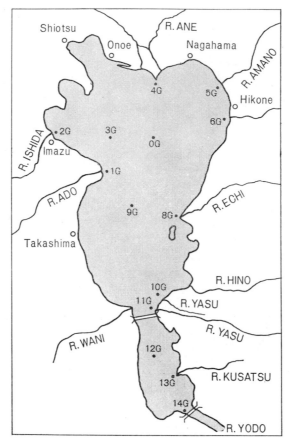

Sampling station	PCB (ng/g, dry basis)	PCB composition
0G	65	KC-400+KC-500 (2 : 1)
1G	25	KC-400+KC-500 (1 : 1)
2G	9	″
3G	50	″
4G	60	KC-400+KC-500 (1 : 2)
5G	15	KC-400+KC-500+KC-600 (1 : 2 : 1)
6G	16	KC-400+KC-500+KC-600 (2 : 2 : 1)
7G	—	
8G	13	KC-400+KC-500+KC-600 (2 : 2 : 1)
9G	26	KC-500
10G	25	KC-400+KC-500 (1 : 1)
11C	23	″
12G	44	KC-500
13G	2700	KC-300+KC-500 (2 : 1)
14G	4400	KC-300+KC-400+KC-500 (1 : 1 : 1)

Fig. 9.9. PCB levels in bottom sediments of Lake Biwa (November, 1972).[13]

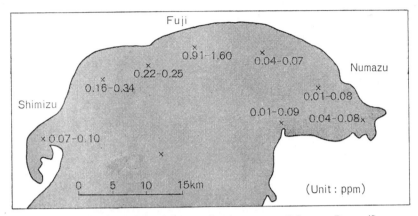

Fig. 9.10. PCB levels in the surface sea water of Suruga Bay. (Source: Shizuoka Prefectural Government; April, 1972)

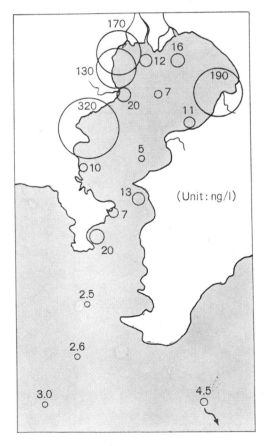

Fig. 9.11. PCB levels in the surface waters of Tokyo Bay (August, 1973).[15]

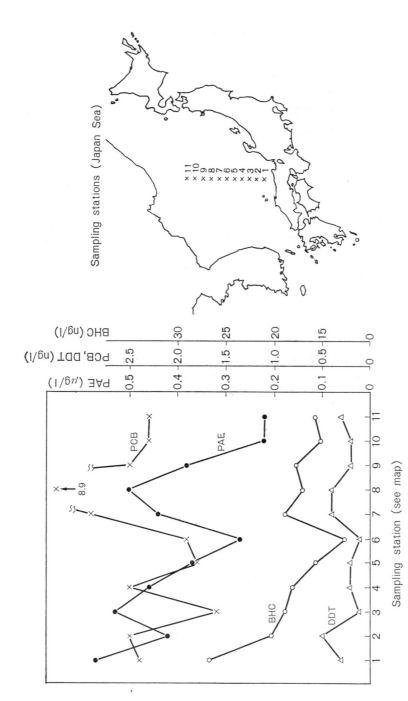

Fig. 9.12. Distribution of some man-made organics in the surface waters of the Japan Sea (July 16–18, 1974).[14]

off Tokyo Bay in August, 1973 (cf. Fig. 9.8).   The high PCB concentrations at this time clearly indicate a continued influx of PCB's into the
environment even after the ban on production.

   The oceanic distribution of PCB's near the Japanese archipelago is
only poorly known, and surveys are now under way.   The results of a
survey on DDT, BHC, PAE and PCB levels in the Japan Sea are given in
Fig. 9.12.[14]   It is interesting to note that concentration peaks occur near
Stations 7 and 8 where the Yamato Bank (a submerged highland) is located and circulating current flows.

   Some examples to show the vertical profiles of organochlorine compounds in the sea are given in Fig. 9.13.[15]   Station 10 is located at the
northern end of Tokyo Bay.   It is interesting to note that β-BHC and
DDT compounds (DDT, DDE and DDD) were detected here only in the
surface waters, a fact which may be explained by the low "solubilities" of
these compounds.   The PCB distribution in the whole profile was rather
uniform.   Stations 45 and 46 are located in the open Pacific, inside and

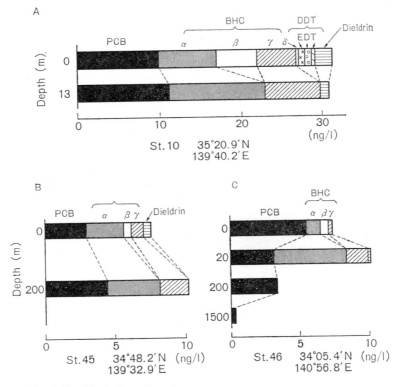

Fig. 9.13. Vertical profiles of some organochlorine compounds in the sea
water of Tokyo Bay (A) and inside and outside the Kuroshio Current (B
and C, respectively) (August, 1973).[15]

just outside the Kuroshio Current, respectively.   (This current is a major current flowing northwards from the tropics to the coast of Japan, and then eastwards towards the North American continent.)   DDT compounds were not detected at Stations 45 and 46 (detection limit, 0.1 ng/l), and β-BHC was found only in the surface waters, as at Station 10.   Outside of the Kuroshio Current, BHC isomers were not detected at a depth of 200 m, in contrast to the situation inside the current.   A possible explanation for this sharp difference is that the Kuroshio Current may act as a "water curtain"; inside the current, pollutants are transported by both the water and the air, but outside the current the main transport route is probably atmospheric.   The PCB distribution at Stations 45 and 46 was rather uniform, and 0.4 ng/l PCB was even detected at a depth of 1500 m at Station 46.

Probable sources of PCB's entering the marine environment (i.e. other than materials of terrestrial origin) are dumping and unintentional leakage from pelagic ships.   This situation, if true, may thus complicate the distribution and fate of PCB's in the marine environment.   Clearly, oceanic surveys of PCB's and other organochlorine compounds are badly needed in order to predict the global fate of PCB's that enter the sea.   These surveys should also examine changes in the compositions of the PCB's as well as the variations in their concentrations.   Moreover, in areas where there is heavy contamination with PCB's the local fish tend to be heavily polluted. The question of pollution of the aquatic and marine environment will thus be discussed again below in the section of fish contamination.

9.3.4   PCB's in the soil and in agricultural products

A nationwide survey[11] of soil samples (88 sites) and unhulled rice samples (37 sites) was undertaken in July through December, 1972 (see Table 9.9).   On the basis of the results, samples showing 10 ppm or more PCB are followed up by intensive investigations of the areas concerned and, depending on these results, appropriate countermeasures are devised. For polluted samples where the PCB level is <10 ppm, the areas are kept under surveillance for future changes.[11]

TABLE 9.9

Results of a Survey on PCB Pollution (Soil and Agricultural Products)[11]

PCB content (ppm)	No. of soil samples	No. of unhulled rice samples
less than 0.01	35 (40%)	26 (79%)
0.01~ 0.10	21 (24%)	5 (15%)
0.11~ 1.0	19 (21%)	1 (3%)
1.1 ~ 10.0	6 (7%)	1 (3%)
10.1 ~100.0	3 (3%)	
more than 100.1	4 (5%)	
Total	88 (100%)	33 (100%)

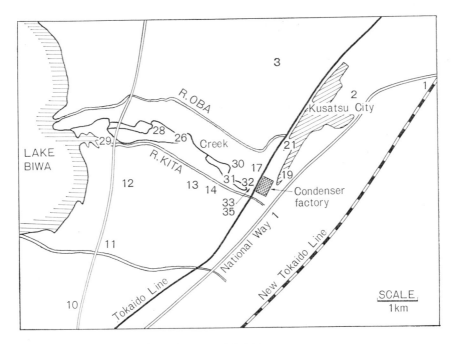

Station†	PCB (μg/g, dry soil basis)
1	0.08
2	0.20
3	0.06
10	0.03
11	0.02
12	0.04
13	0.19
14	0.39
17	0.40
19	0.43
21	0.38
26	5.1
28	6.1
29	8.7
30	86
31	55
32	10,000
33	3200
35	6800

† See map.

Fig. 9.14. PCB levels in soil samples near a condenser factory, Kusatsu City (April, 1972)[16]

TABLE 9.10
Experimental uptake of PCB's from Contaminated Soil by Rice Plants[17]

PCB added to soil (ppm)	PCB level	
	Straw (ppm)	Unhulled rice (ppm)
Water-logged conditions		
0	0.15	0.02
10	0.16	0.04
100	0.79	0.02
1000	4.08	0.03
Kusatsu soil†	0.90	0.03
Upland conditions		
0	0.12	0.02
10	0.16	0.02
100	0.34	0.02
1000	0.73	0.02
Kusatsu soil†	0.35	0.02

† PCB-contaminated soil near waste-water from a condenser factory (PCB concentration, 24.9 ppm; mainly KC-300).

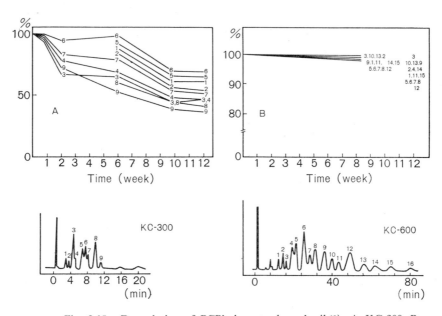

Fig. 9.15. Degradation of PCB's in water-logged soil.[18] A, KC-300; B, KC-600. The numerals on the degradation curves refer to the numbered peaks indicated on the respective gas chromatograms.

Fig. 9.14[16]) shows the PCB levels found in soil samples near a condenser factory in Kusatsu City (see Fig. 9.3, no. 25) and indicates the passage of PCB's through the irrigation creeks that flow away from the factory. Similar results have been obtained in a survey conducted by the Shiga prefectural authorities, who detected a maximum concentration of 1.33 ppm PCB in unhulled rice.

Masujima et al.[17]) have investigated the uptake of PCB's by rice plants grown in pots containing PCB-contaminated soil. Although the PCB concentrations in the rice straw showed a general (but not proportional) increase with increasing amount of PCB's added to the soil, the PCB concentrations in unhulled rice were rather uniform throughout the four test levels of PCB's (see Table 9.10). This experiment suggests that there may be no or only very slow transport of PCB's through rice plants, and that the high concentration in unhulled rice detected in Shiga Pref. could be due to the adsorption of atmospheric PCB's onto the rice tissue surface.

The degradation of low chlorine PCB's (KC-300 and KC-400) is known to proceed slowly in paddy soil that lies in a water-logged condition after a period of drying, but high chlorine PCB's (KC-500 and KC-600) are very stable under these conditions (see Fig. 9.15).[18]) Also, under identical conditions, about 90% of DDT and BHC isomers are degraded within a week.[19]) Evaporation of PCB's from the soil surface during the dry season (fall to spring) may be one major route for PCB loss from the soil. However, in permanently submerged marine sediments (where the bulk of the environmental PCB load accumulates), such degradation and/or PCB loss by evaporation cannot be expected.

9.3.5 PCB's in fish

Since the bulk of the environmental PCB load exists in the marine environment, the marine biota will undoubtedly remain affected for a long time to come. Moreover, fish represent a major source of protein for the Japanese people. The severe pollution of fish caused by PCB's has already led to considerable social unrest, and detailed investigations have been undertaken by several laboratories.

A nationwide survey[11]) was carried out in May through December, 1972, by the Fisheries Agency and Environment Agency; it covered 110 selected areas and a total of 599 samples (Table 9.11). A contamination level of >1 ppm PCB was observed in the fresh edible parts of 16% of sea-water fish and 18% of fresh-water fish.

Combining the results of Government surveys in 1972[20]) and 1973,[21]) Doguchi[22]) compiled the data summary given in Fig. 9.16 and Table 9.12. As shown in the figure, the PCB pollution of the aquatic environment (as represented by the levels in fish) is lower in the northern part of Japan

TABLE 9.11
Results of a Survey on PCB Pollution (Fisheries Products)[11]

Material \ PCB content (ppm)	0.09 or less	0.1~0.4	0.5~0.9	1~2	3	more than 3	Total
Sea-water fish	(34) 155	(38) 175	(12) 53	(11) 48	(2) 11	(3) 14	(100) 456
Fresh-water fish	(15) 21	(44) 63	(23) 33	(13) 19	(1) 2	(4) 5	(100) 143
Total	(30) 176	(40) 238	(14) 86	(11) 67	(2) 13	(3) 19	(100) 599

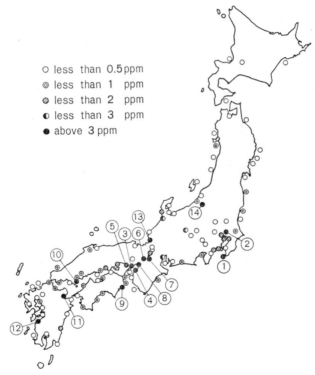

○ less than 0.5 ppm
◎ less than 1 ppm
⦸ less than 2 ppm
◐ less than 3 ppm
● above 3 ppm

Fig. 9.16.   PCB residues in fish.[20-22]   The numbered locations indicate the areas referred to in Table 9.12.

and higher in southwestern areas.   Details of the highly-polluted coastal and inland-water areas where the PCB level was >3 ppm on the basis of fresh edible parts, are summarized in Table 9.12.   The most polluted localities were as follows: south of Lake Biwa and the River Seta, Iwakuni, Tsuruga Bay, Tokyo Bay, Osaka Bay, and Harimanada.

The wide variety of PCB compositions found in fish (see Fig. 9.17)

Table 9.12

TABLE 9.12

Highly Polluted Coastal and Inland-water Areas where PCB Levels
Higher than 3 ppm Were Detected in Fish[20-22]

Area[1]	Total no. of samples	A[2]	Species of fish in which PCB levels higher than 3 ppm were detected[4]		
			Species	PCB range (ppm)	B[3]
(1) Tateyama	5	1/5	*Mugil cephalus* (grey mullet)	0.9 – 18	1/5
(2) Middle reaches of River Arakawa	6	3/6	*Carassius carassius* (crucian carp)	6 – 21	2/2
			Zacco platpus (minnow)	9.2 – 10	1/2
(3) Northwest of Osaka Bay	160	20/160	*Clupanodon punctatus*	4 – 6	10/10
			Scomber japonicus (mackerel)	4 – 7	10/10
(4) South of Osaka Bay	6	1/6	*Scomber japonicus*	2 – 5	1/2
(5) Harimanada	160	42/160	*Mugil cephalus*	2 – 8	12/15
			Clupanodon punctatus	2 – 44	10/15
			Lateolabrax japonicus (sea bass)	2 – 24	10/15
			Clupanodon punctatus	4 – 5	5/5
			Lateolabrax japonicus	4 – 22	5/5
(6) River Yodo and Yura	160	15/160	*Zacco platypus*	3 – 4	5/10
			Carassius carassius	0.6 – 7	1/10
			Zacco platypus	3 – 5	6/10
			Carassius carassius	1 – 5	3/10
(7) South of Lake Biwa and River Seta	320	19/320	*Carassius carassius*	0.4 – 6	3/10
			Parasilurus asotus (catfish)	2 – 18	5/7
			Zacco platypus	4 – 6	10/10
			Carassius carassius	1 – 5	1/10
(8) East of Osaka Bay	160	1/160	*Trichiurus lepturus*	0.6 – 4	1/10
(9) Port of Hiwasa	5	1/5	*Mugil cephalus*	0.3 – 5	1/5
(10) Iwakuni	160	23/160	*Mugil cephalus*	0.3 – 79	7/10
			Mugil cephalus	0.2 – 40	4/10
			Clupanodon punctatus	1 – 5	3/10
			Ditrema temmincki	0.8 – 7	4/10
			Conger Myriaster (congereel)	5 – 10	5/5
(11) Beppu Bay	160	8/160	*Anguilla japonica* (eel)	0.03–130	8/19
(12) Minamata Bay	5	1/5	*Mugil cephalus*	0.01– 7	1/5
(13) Tsuruga Bay	159	14/159	*Lateolabrax japonicus*	0.1 – 10	2/6
			Lateolabrax japonicus	2 – 18	1/2
			Fishes in the port	0.2 –110	10/15
(14) River Seki	80	11/80	*Tribolodon hakonensis*	0.1 – 17	5/10
			Tribolodon hakonensis	1 1 7	2/10
			Tribolodon hakonensis	0.2 – 4	1/10
Estuary of River Seki	5	1/5	*Kareius bicoloratus* (flatfish)	2 – 4	1/5

[1] See Fig. 9.16.

[2] A: no. of samples in which PCB level exceeds 3 ppm/total no. of samples.

[3] B: no. of samples in which PCB level exceeds 3 ppm/no. of samples of each species.

[4] All PCB levels on fresh tissue basis.

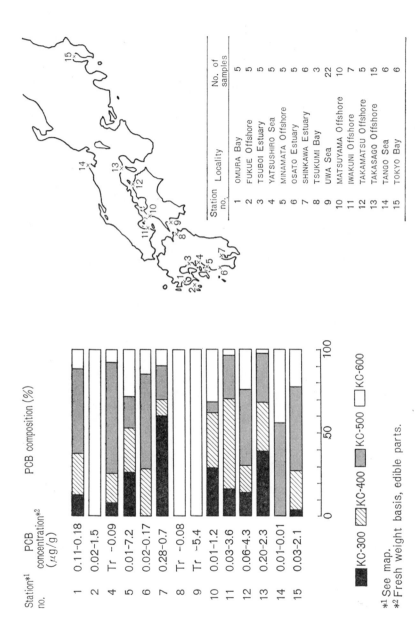

Fig. 9.17.   PCB compositions in fish.[14]

appears to reflect the diversity of PCB consumption and discharge into marine waters and sediments in different areas.   This is apparently due in part to the rather weak activity of fish for metabolizing PCB's, so that the PCB composition in the aquatic environment is generally reflected in the PCB composition in the fish.   The actual PCB levels in fish vary with the species, age, body weight, kinds of tissues and fat content, and appear to show seasonal fluctuations.   Similarly to other organochlorine pollutants, tissues of high fat content tend to show high PCB levels, and the stabler high chlorine PCB's tend to accumulate more readily than low chlorine PCB's.   For example, in yearling carp (*Cyprinus carpis*) fed on PCB-containing foods for 12 months, it was found[23] that 22.5, 37.0 and 72.3% of the KC-300, KC-400 and KC-500, respectively, persisted and accumulated in the fish (Table 9.13).   Moreover, the gas-chromatographic pattern of the accumulated KC-300 was considerably different from that of the original KC-300, but was less changed in the case of KC-400 and almost unchanged in the case of KC-500.

Concentration factors (field observation ratios) for PCB's and certain other man-made organic pollutants are shown in Table 9.14.[14]   These concentration factors (C.F.) were in the order PCB>DDT>dieldrin> BHC>phthalic acid ether (DEHP), and C.F. for PCB, being of the order of 10^5, is the highest among the investigated man-made organics.   This field survey also indicated that the lower the solubility (in water) of the organics, the higher was the C.F. value.   This inverse relationship appears to suggest that direct PCB uptake from the water is a major route for PCB uptake by fish.

TABLE 9.13

PCB Accumulation in Fish Fed on PCB-containing Feed for 12 Months[23]

PCB	Daily dose (mg/kg)	Rate of accumulation (%)†
KC-300	0.127	22.5
KC-400	0.121	37.0
KC-500	0.134	72.3

† (PCB in fish÷total dose) × 100.

TABLE 9.14

Concentration Factors (Field Observation Ratios) of Some Man-made Organics for Striped Mullets in the Seto Inland Sea (1972–1974)[14]

		PCB	α-BHC	β-BHC	γ-BHC	δ-BHC	T-BHC
C. F. (F/W)	Max.	680,000	4,400	8,900	3,300	750	7,900
	Min.	2,300	70	43	39	50	86
	Ave.	77,700	780	910	910	630	840

		DDE	DDD	DDT	T-DDT	Dieldrin	DEHP
C. F. (F/W)	Max.	340,000	99,000	260,000	550,000	8,400	54
	Min.	1,900	1,700	1,200	1,700	180	23
	Ave.	43,000	17,000	62,000	57,000	2,500	42

No. of Samples: 62. F: Concentration in Fish Meat. W: Concentration in Sea Water.

9.3.6 PCB's in birds

High levels of PCB's have been detected in wild birds from the Kanto Plain and Tokyo Bay (see Table 9.15).[22]   There was considerable variation from specimen to specimen, particularly in the case of the aigrette (*Egretta garzetta*), where the lowest level in muscle was 0.3 ppm and the highest 180 ppm PCB.

The results of a survey on PCB's and other organochlorine compounds in wild birds conducted in the Matsuyama Plain during 1972–73 are listed in Tables 9.16 and 9.17.[24]   The number of species investigated was 10 (representing different habitats and food intakes) and the number of sample birds was 95.   All were collected by shooting, and PCB's were found in all samples.   The PCB levels were high in the fish-eating black-tailed gull (*Larus crassirostris*), carnivorous black-eared kite (*Milvus migrans lineatus*) and omnivorous crow (*corvus* sp.), but were low in herbivorous birds such as the brown-eared bulbul (*Hypsipetes amaurotis amaurotis*) and dusky thrush (*Turdus naumanni eunomus*).

TABLE 9.15

PCB Residues in Egrets Collected from 3 Colonies in the Kanto Plain and Gulls from Tokyo Bay[22]

Species		Breast muscle			Liver			Date	History
		Fat content (%)	ppm wet basis	ppm fat basis	Fat content (%)	ppm wet basis	ppm fat basis		
		Adult birds							
	1	—	—	—	2.4	4	170	Aug. 1971	Dying
	2	3.8	18	470	6.3	12	190		Dead
	3	2.8	6	210	3.5	7	200		Dead
	4	—	2	—	—	6	—	June 1971	Dead
	5	7.1	51	720	4.8	108	2300		Dead
	6	1.1	180	16,000	—	—	—		Dying
	7	—	2	—	—	2	—	Nov. 1971	Dead
	8	2.2	45	2100	3.7	52	1400		Dead
	9	1.1	6	500	3.0	23	780		Dead
	10	1.1	0.3	22	2.2	1.4	60		Dying
	11	0.4	0.5	110	3.4	3	90		Dying
Aigrette[†1]		*Juvenile birds*							
	12	0.7	1	140	3.6	5	130	June 1972	Healthy
	13	3.2	1.3	40	4.1	0.8	20		Healthy
	14	0.8	0.05	9	4.7	0.3	5		Healthy
	15	1.1	1.1	90	2.5	3	130		Healthy
	16	1.0	0.1	13	4.3	0.01	0.2		Healthy
	17	1.9	5	260	3.2	18	550		Healthy
		Eggs							
		5.2	5	90					
		7.4	3	40					

†1 *Egretta garzetta.*

TABLE 9.15—*Continued*

Species		Breast muscle			Liver			Date	Date
		Fat content (%)	ppm wet basis	ppm fat basis	Fat content (%)	ppm wet (%)	ppm fat basis		
Great white egret (nestling) [2]		0.9	3	300	3.3	20	610	Aug. 1972	Dead
		Adult brids							
Black-tailed gull[3]	1	8.1	6	74	4.9	2	41	⎫	Shot
	2	—	39	—	—	3	—	⎪	Shot
	3	5.1	3	59	6.7	45	670	⎪	Shot
	4	—	16	—	—	13	—	⎬ Nov. 1971	Shot
	5	2.4	21	880	4.0	18	450	⎪	shot
	6	4.8	17	350	4.8	17	350	⎪	Shot
	7	5.5	26	470	6.4	16	250	⎪	Shot
	8	—	9	—	—	5	—	⎭	Shot
Kamchatkan black-headed gull[4]		—	2	—	—	2	—	Nov. 1971	Shot

[2] *Egretta albamodesta.*

[3] *Larus crassirostris.*

[4] *Larus ridibundus.*

An analysis of formalin-preserved specimens of the magpie (*Pica pica*) from the past 16 yr has suggested the persistence of PCB, DDT and BHC residues in this bird (see Fig. 9.18).[15]   However, after the ban on the use of BHC and DDT, the levels of these two pollutants (particularly BHC) dropped sharply.   On the other hand, the PCB level in *P. pica* was still high in 1973, and there was no declining tendency.   This result may be

TABLE 9.16

Distribution of Residue Levels[1] and Number of Detections of Some Man-made Organochlorine Compounds in the Breast Muscle of Wild Birds[2] (Matsuyama Plain, 1972-73)[24]

	No.	%	≥1.00 ppm No.	≥1.00 ppm %	≥0.10 ppm No.	≥0.10 ppm %	≥0.01 ppm No.	≥0.01 ppm %	≥0.001 ppm No.	≥0.001 ppm %	<0.001 ppm No.	<0.001 ppm %
PCB	95	100	24	25	13	14	23	24	—	—	35	37
Total-BHC	95	100	2	2	23	24	64	67	6	7	—	—
α-	95	100	—	—	—	—	26	27	58	61	11	12
β-	93	98	2	2	19	20	60	63	10	11	4	4
γ-	95	100	—	—	—	—	15	16	56	59	24	25
δ-	68	72	—	—	—	—	—	—	51	54	44	46
DDT compounds	95	100	2	2	40	42	47	50	6	6	—	—
o,p'-DDE	44	46	—	—	—	—	4	4	26	28	65	68
p,p'-DDE	95	100	1	1	35	37	48	50	10	11	1	1
p,p'-DDD	60	63	1	1	16	17	18	19	20	21	40	42
o,p'-DDT	35	37	—	—	—	—	7	7	14	15	74	78
p,p'-DDT	53	56	—	—	—	—	19	20	27	28	49	52
Dieldrin	64	67	—	—	6	6	32	34	21	22	36	38

[1] Unit: $\mu g/g$, on fresh tissue basis.

[2] No. of samples analyzed=95.

Table 9.17

Organochlorine Residues[†1] in the Breast Muscle of Wild Birds (Matsuyama Plain, 1972–73)[24]

Species	No.	PCB			BHC[†2]			DDT[†3]			Dieldrin		
		max.	min.	av.	max.	min.	av.	max.	min.	av.	max.	min.	av.
Black-tailed gull[†4]	10	8.04	1.62	3.71	0.16	0.03	0.08	0.73	0.11	0.38	0.94	0.01	0.03
Black-eared kite[†5]	10	3.57	0.86	2.40	1.27	0.10	0.48	2.51	0.41	0.82	0.28	0.04	0.11
Japanese jungle crow[†6] Eastern carrion crow[†7]	10	1.20	0.05	0.28	0.32	0.04	0.10	1.01	0.03	0.24	0.43	ND	0.05
Brown-eared bulbul[†8]	9	4.18	TR	0.03	0.14	0.01	0.04	0.05	0.02	0.03	0.00	ND	0.00
Dusky thrush[†9]	9	0.05	TR	0.03	0.12	0.01	0.04	0.05	0.02	0.03	0.00	0.00	0.00

[†1] Unit: μg/g, on fresh tissue basis.

[†2] Sum of 4 isomers (α, β, γ and δ)

[†3] Sum of p,p'-DDT, p,p'-DDE, p,p'-DDD, o,p'-DDT and o,p'-DDE.

[†4] Larus crassirostris.   [†5] Milvus migrans lineatus.   [†6] Corvus levaillantii japonensis.

[†7] Corvus corone orientallis.   [†8] Hypsipetes amaurotis amaurotis.

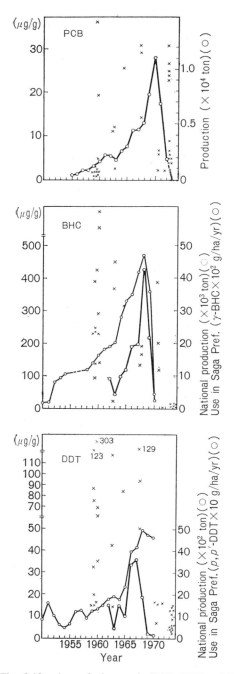

Fig. 9.18. Annual changes in PCB, BHC and DDT residues in the breast muscle of the magpie, *Pica pica* (Saga Plain), and annual productions (●) and use (○) of PCB, BHC and DDT.[15] ×, Residue values (on fat basis) of PCB, BHC and DDT.

explained by both the high persistence of PCB's and by a continuing input of PCB's into the natural environment.

9.3.7 PCB uptake by man

Judging from the PCB residue levels in drinking water and the atmosphere, almost all of the human uptake of PCB's can be assumed to be derived from the diet, except in workers in PCB-handling factories, etc. However, the daily amounts of PCB intake by the Japanese people are not known in detail.

As an example, the results of a survey conducted in Matsuyama City (see Fig. 9.3, no. 38) are shown in Table 9.18.[25] This work was under-

TABLE 9.18

Amounts of PCB's[†1] Taken from Meals (August–September, 1972, Matsuyama City)[25]

Food group \ Day	1	2	3	4	5	6	7	Av.
1) Cereals	0.3	1	Tr[†2]	0.1	Tr	0.1	Tr	0.2
2) Potatoes	Tr	Tr	—[†3]				0.1	Tr
3) Sugars		—	0.1			—		Tr
4) Fats and oils	0.1	0.1	0.1	0.1	—	—		Tr
5) Soybeans and soybean products	1.3	—	0.6		—	—	—	0.2
6) Fruits	—	Tr	—	0.2	Tr	—	—	Tr
7) Vegetables	0.6	0.2	0.1	0.6	Tr	—	0.6	0.3
8) Fish and other sea-foods	47	17	0.8	17	3.6	0.5	3.7	13
9) Meat		0.5	3.3	0.8	Tr	1	1.2	1
10) Eggs	0.3	—	1.5	0.7	2.3	1.6		0.9
12) Milk	0.4	0.6	0.4	0.4	0.4	0.6	0.5	0.5
13) Others		—		0.1	0.5		0.8	0.2
Uptake of PCB's	50	19	6.9	20	6.8	3.8	6.9	16

†1 Unit: μg.   †2 Trace.   †3 None detected.

taken as a survey of the PCB content of complete, prepared meals. The foods were bought at markets in Matsuyama City in August and September, 1972, and were cooked and prepared according to a prescribed menu for one week. The total amounts of PCB's contained in the one-day diets varied in the range 3.8–50 μg per capita, with a daily average of 16 μg per capita. The food material having the highest PCB content was fish (especially that derived from the Seto Inland Sea), and of the PCB present in the one-week diets, 80% occurred in fish, 6% in meat, and 6% in eggs. The amounts of PCB's in vegetable foods were found to be small or undetectable. The ratios of different PCB homologs in the foods were as follows: KC-300, 30%; KC-400, 4%; KC-500, 23%; KC-600, 43%.

9.3.8   PCB's in man

PCB and pesticide levels that have been reported from human fat in various parts of Japan[14,26-28] are listed in Table 9.19.   As can be seen, the PCB levels were generally lower than those of BHC and DDT compounds, but they were still considerably high.   It is noteworthy, also, that the highest PCB levels occurred in fishermen, presumably as a result of their consistently eating PCB-contaminated fish.   Similar to the case of BHC and DDT, high fat tissues generally showed high PCB levels, and the

TABLE 9.19
PCB and Pesticide Residues[†1] in Human Fat

Area	Period	No. of samples	PCB mean	PCB range	β-BHC	Total DDT	Dieldrin	Ref.
Kochi (39) [†2] 1971–72	1971–72	50	2.86	0.61–18.04	7.87	5.44	0.31	26
male		31	3.71		8.62	5.51	0.26	
female		19	1.48		6.66	5.33	0.22	
farmer		7	2.75		7.87	5.24	0.31	
fisherman		5	7.50		6.95	7.79	0.19	
clerk		6	2.60		7.26	4.97	0.21	
others		11	2.92		8.52	5.41	0.22	
jobless		21	1.89		7.93	5.10	0.25	
Kyoto (26)	1971	10	4.7	1.9 –13.3	10.9	9.8	0.19	27
Tokyo (13)	1971	40	1.85	0.1 –10	5.16	4.19	0.59	28
Tokyo (13)	1971–72	27	2.1	0.7 – 5.1				22
Ehime (38)	1971	16	1	0.1 – 2				28

[†1] Unit: μg/g, on fat basis.
[†2] ( ): See Fig. 9.3

homologs occurring in human fat are mostly high chlorine PCB's, viz. KC-500 and KC-600.
Nationwide surveys of the PCB content of human milk were conducted by the Ministry of Health and Welfare in 1972, 1973, and 1974, as summarized in Fig. 9.19.   Geographically, the western part of Japan (particularly the coast of the Seto Inland Sea) shows high PCB levels in human milk.   Although a gradual overall decrease can be distinguished over the period 1972–74, one quarter of all samples still exceeds a PCB content of 0.033 ppm in 1974.   (This level represents the tolerance limit calculated in Japan for human milk based on an "acceptable daily intake" of 5 μg/kg body wt and a daily intake of milk by the babies of 150 ml/kg body weight.)

Fig. 9.19. (on page 178)   PCB levels in human milk in Japan (1972–74). Unit: ppm, on a whole milk basis. (Source: Ministry of Health and Welfare, 1973–75)

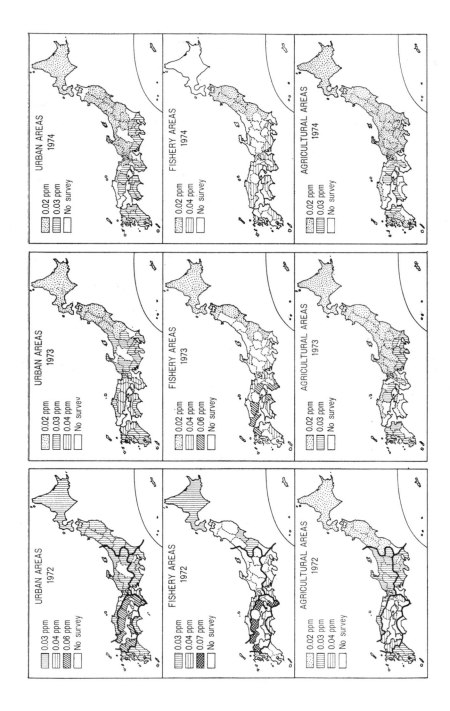

REFERENCES

1. T. Mizutani, M. Matsumoto and K. Fujiwara, *J. Food Hyg. Soc. Japan* (Japanese), **12**, 232 (1971).
2. R. Tatsukawa, *J. Environ. Poll. Control* (Japanese), **7**, 419 (1971).
3. T. Wakimoto, R. Tatsukawa and T. Ogawa, *ibid.*, **7**, 517 (1971).
4. N. Isono, *Chemicals and Man* (Japanese), pp. 216, Chuokoronsha, 1975.
5. N. Isono, *PCB, Its Past and Present* (Japanese), pp. 52, 1973.
6. R. Tatsukawa, *PCB* (Japanese), p. 170, Asahi Shinbunsha, 1972.
7. I. Hara, *Repts. Osaka Inst. Public Health, Ind. Hyg.*, Section no. 7, 26 (1969).
8. T. Wakimoto, R. Tatsukawa, T. Ogawa and I. Watanabe, *Japan Analyst* (Japanese), **23**, 790 (1974).
9. R. Tatsukawa and I. Watanabe, *Science of Food* (Japanese), no. 8, 49 (1972).
10. R. Tatsukawa, *Japan, J. Ecol.* (Japanese), **23**, 74 (1973).
11. Environment Agency, *Quality of the Environment of Japan, 1973*, 1973.
12. R. Tatsukawa and I. Watanabe, *Science of Food* (Japanese), no. 8, 55 (1972).
13. R. Tatsukawa and I. Watanabe, presented at the 38th Meeting of the Japanese Society of Limnology, Matsuyama, 1973.
14. R. Tatsukawa and S. Tanabe, *unpublished data.*
15. M. Fukushima, *M.S. Thesis, College of Agriculture* (Japanese), *Ehime Univ.*, 1974.
16. R. Tatsukawa, I. Watanabe and T. Wakimoto, *unpublished data.*
17. Environment Agency, *Comprehensive Research Report on PCB Pollution and Its Control Measures* (Japanese), p. 435, 1973.
18. T. Wakimoto, T. Yakushiji and R. Tatsukawa, presented at the Meeting of the Society of the Science of Soil and Manure, Japan, 1971.
19. R. Tatsukawa, T. Wakimoto and T. Ogawa, *J. Food Hyg. Soc. Japan* (Japanese), **11**, 1 (1970).
20. Environment Agency and Fisheries Agency of Japan, *Report on PCB Pollution of Fish* (Japanese), December, 1972.
21. Fisheries Agency of Japan, *Progressive Report on PCB Pollution of Fish* (Japanese), June, 1973.
22. M. Doguchi, *Collection of Papers* presented at the Research Conference on New Methodology in Ecological Chemistry, Susono, Japan, November, 1973, p. 269, 1973.
23. A. Wakabayashi *et al.*, *Ann. Rept. Tokyo Metrop. Res. Inst. Environ. Protect.* **5**, 80 (1974).
24. R. Tatsukawa, *Occurrence and Distribution of Environ. Pollutants in Wild Birds, Ehime Pref., Japan, (I) Organochlorine Compounds* (Japanese), pp. 52, 1973.
25. R. Tatsukawa, K. Kawaita and T. Wakimoto, *J. Home Econ.* (Japanese), **24**, 466 (1973).
26. T. Nishimoto, M. Ueta and S. Taue, *Igaku no Ayumi* (Japanese), **82**, 515 (1972).
27. T. Mizutani, M. Matsumoto and K. Fujisawa, *ibid.*, **81**, 666 (1972).
28. R. Tatsukawa and K. Kubo, *unpublished data.*

Index